Electron-Gated
Ion Channels

Electron-Gated Ion Channels

With Amplification by NH$_3$ Inversion Resonance

Wilson P. Ralston

BioKinetix Research
Stamford, Connecticut

SCITECH
PUBLISHING, INC.
Raleigh, NC

ELECTRON-GATED ION CHANNELS

SciTech Publishing, Inc.
911 Paverstone Drive, Suite B
Raleigh, NC 27615
Email: info@scitechpub.com
www.scitechpub.com

10 9 8 7 6 5 4 3 2 1

Library of Congress Cataloging-in-Publication Data

Ralston, Wilson P.
    Electron-gated ion channels : with amplification by NH3 inversion
    resonance / Wilson P. Ralston.
           p. cm.
    Includes bibliographical references and index.
    ISBN 1-891121-41-3
    1. Ion channels. 2. Ion channels–Mathematical models.
    3. Tunneling (Physics). I. Title.
QH603.I54 R35 2005
571.6'4–dc22   2005000102

To Luitgard

CONTENTS

PREFACE

Ion channels play a central role in biological signaling and the gating mechanism for controlling ion conduction through a channel is of fundamental importance to cellular function and our nervous system. When I first started investigating the gating mechanism, it was with a conviction that electron tunneling must somehow be involved. I had noted a similarity between the sodium ion channel and a two-terminal amplifying device, the tunnel diode. When the negative conductance I–V curve for the tunnel diode (rotated 180° with a voltage offset) was compared with the negative conductance I–V curve for the sodium channel, there was a striking resemblance. There had also been a number of published experimental studies of electron tunneling across proteins and across amino acids in the α-helix configuration. If electron tunneling is occurring across amino acids in an α-helix, I thought, why shouldn't it be occurring across the transmembrane α-helix segments of ion channel proteins? Could the negative conductance region of the sodium channel I–V curve be due to electron tunneling? These thoughts eventually led me to develop a model for ion channel gating, based upon amplified resonant electron tunneling.

After becoming acquainted with the ion channel equations of Hodgkin and Huxley, I started developing equations for electron tunneling with the aid of math software, focusing on the gating of the sodium channel. If electron tunneling was controlling gating, the peak electron-tunneling time constant, and the slope sensitivity for electron probability, should be capable of matching the time constant and slope sensitivity determined by Hodgkin and Huxley for the sodium channel of the squid giant axon.

It also became clear that if electron tunneling was occurring across ion channel proteins, it most likely would use the arginine and lysine amino acids of the S4 region. This region is referred to in the literature as the "voltage sensor" and is thought to be associated with gating. From cDNA and the amino acid sequence it had been determined that every third residue of the voltage sensor was arginine or lysine. This spacing gave a rise distance of 4.5 Å along the α-helix axis and a calculated diagonal center-to-center tunneling distance of about 6 Å. Using this tunneling distance in my equations, the peak time constant could match that for the sodium channel, but the slope sensitivity was more than an order of magnitude too low. It became clear that an electron-tunneling model for gating would not be possible, unless a means for amplification could be found.

Polar molecules can stretch in an electric field, and I wondered if a change in membrane voltage that caused stretching, thereby changing the tunneling distance, might provide the needed amplification. An equation for amplification based on Hooke's Law was developed using a spring model to represent molecular stretching. Combining this with the electron tunneling equation did provide amplification and allowed a match with the sodium channel slope sensitivity. The problem now was, that to obtain a matching slope sensitivity, the force constant of the spring had to be very low – about three orders of magnitude lower than the usual force constant of molecules, which have stretching frequencies in the infrared region.

The stretching frequency for the spring model was calculated to be 21 GHz. This was very close to the well-known inversion frequency of NH_3, which is about 24 GHz in the gas phase. The NH_3 inversion frequency had been used in the first atomic (molecular) clock and the first molecular amplifier, the ammonia maser.

I became convinced that the NH_3 group at the end of the arginine and lysine side chains was inverting and that this was the basis for the amplification. In an aqueous environment, this group is ionized, carrying a net positive charge, and depending on pH, may be NH_2^+ or NH_3^+. However, the inversion is unlikely to be taking place in an aqueous environment, due to energy loss to the surrounding water molecules.

The working hypothesis was that arginine and lysine amino acids of the S4 region had NH_3 side chain groups that were protected from contact with water molecules by the surrounding transmembrane α-helix structures, and that they did not carry a net excess charge, except when a shared electron tunneled to them. An additional element of the hypothesis was that the bonding to carbon and the rest of the side chain did not interfere with the NH_3 inversion resonance, but simply restricted the motion of the nitrogen atom by increasing its effective mass. This would lower the inversion frequency from the gas-phase value.

In order to confirm this working hypothesis, I looked for a way to measure the microwave frequency that had been calculated. I wanted to preserve the native state of arginine and lysine in the protein during measurement. Microwave spectroscopy is normally restricted to the gas phase because strong water absorption bands mask the sample absorbance in the liquid phase. Could there be a way around this problem? What if fluorescence was used instead?

When microwaves are absorbed by the sample, there is a temperature rise. Most fluorescent molecules have a small temperature coefficient and I wondered if, by using temperature-sensitive fluorescence detection (optimized to measure a small signal on top of a big one), the microwave absorption of a sample might be detected with less interference from water. To detect the temperature change of NH_3 side chain groups, the fluorophore would have to be close to the arginine or lysine residues and protected from intimate contact with the surrounding water molecules. To conduct the experiment, a new type spectroscopy instrument, for recording protein spectra using thermally modulated fluorescence, would have to be constructed and a microwave swept signal generator and amplifier would need to be obtained. It all seemed like a long shot, but I thought it would be worth trying. The new spectroscopy technique and the experimental observations resulting from it are described in Part II of this book.

The sample chosen for the experiments was Blue Fluorescent Protein (BFP), which contained arginine and lysine residues in a cage that protected key residues and the fluorophore from water. The frequency scans for BFP showed numerous peaks in the 11.8 to 20 GHz region that were not due to water. Nearly all of the peaks in this region matched the strongest NH_3 gas-phase inversion lines scaled down with two different scaling factors. The two sets of microwave frequencies corresponded to two inversion resonances, one at 14.3 GHz and the other at 16.8 GHz.

This unexpected finding of two sets of scaled frequencies led to the hypothesis that arginine has two NH_3 side chain groups undergoing inversion; one at 16.8 GHz, with a single bond between nitrogen and carbon, and the other at 14.3 GHz with a more restrictive double bond to carbon. The lysine NH_3 group, having a single bond to carbon, would have the higher frequency. After these findings, adjustments were made to the model to match the new frequencies while retaining agreement with the Hodgkin-Huxley slope sensitivity and time constant.

Introducing the electron into gating brought many constraints. Obtaining solutions in one area would often set criteria in other areas, sometimes requiring a revision to what was previously done. This interlinking and rigidity was essential in defining the final system. Research findings are usually published in journals, but for 150-pages of interlinking material, with many equations, this was not an option, hence this book.

The theory for amplified electron tunneling and gating was developed based on sodium and potassium channels, but it also applies to many other channel types. A model for calcium oscillators was developed as an application of the theory.

The models and theory rely on the experiments and published work of many researchers. The classic work of Hodgkin and Huxley, and more recent ion channel studies, and also, electron tunneling studies, and amino acid sequence data were invaluable in shaping the theory and keeping it in agreement with experimental observations. And finally, the alignment of the numerous frequencies for Blue Fluorescent Protein with the scaled rotational-vibrational frequencies for gas-phase NH_3 gave a confirmation that NH_3 was indeed inverting and that this was a likely mechanism for amplification.

The focus of the book is on the developments related to amplified electron tunneling and gating. A familiarity with ion channels and with mathematics through calculus is assumed. An effort was made to keep the equations and their derivations simple by making, where appropriate, first-order approximations and by using circuit analogies. For further information on ion channels, electron tunneling, or microwave spectroscopy, the reader is referred to the reference section.

Wilson Ralston

Stamford, Connecticut
November 2004

Chapter 1

INTRODUCTION

Ion channel gating influences many cellular processes. It is at the center of electrical and calcium signaling in nerve cells and is directly responsible for the triggering of action potentials. The frequency of action potential pulses encodes sensory information, which is transmitted via axons throughout the nervous system. The axon uses principally sodium and potassium ion channels to transmit voltage pulses of about one-tenth volt, while maintaining constant pulse amplitude over distances up to one meter. This is possible because amplification, occurring in the ion channel protein, gives the sodium channel a negative resistance (negative conductance) characteristic. Negative resistance allows sodium channels to generate current pulses, which supply energy to the pulses propagating through the axon, thus compensating for positive resistance energy loss in the axon.

The gating of ion channels has been intensively investigated over the last 10 to 15 years with both experimental and theoretical methods. Most current models for gating involve the movement of the S4 voltage sensor and the positively charged residues, arginine and lysine. An early model for gating was proposed by Armstrong and Bezanilla (1977) to explain the fast sodium channel inactivation. It became known as the "ball and chain" model. More recently it has been used to explain the fast inactivation in potassium channels, particularly in *Shaker* channels.

The quantum-mechanical approach to channel gating described in this book is based on single-electron tunneling across arginine and lysine residues of the S4 transmembrane protein segment (Fig. 1-1). Models for controlling the gating of ion channels by electrons have most likely been considered by other researchers; however, there is a problem – an electron-gating model requires a mechanism for amplification in order to match the experimental data for ion channel voltage sensitivity. Amplification based on the inversion of NH_3, addresses this sensitivity problem.

The quantum-mechanical approach to gating has a number of advantages over previous gating models. It is intrinsically simple; the electric field from an electron closes a gate by increasing the height of an energy barrier at a gating cavity in the channel. The energy change for a Na^+ channel gating barrier was calculated as 180 meV (Section 5-1). Adding 50 meV for the open-gate energy barrier (Section 5-3) gives a total energy barrier of 230 meV (9.5RT), in agreement with the Na^+ channel energy barrier values determined by Hille (1975).

Another advantage over previous gating models is that with amplified electron gating, the energy required to open or close a gate is very small. It is well below the thermal noise energy (kT). The thermal noise has a flat frequency distribution with random impulses in the time domain, and this permits differentiation between the noise (N) and the multichannel-correlated voltage-dependent gating signal (S). Based on established theory, the S/N for N' channels would increase with the square root of N'.

Electron gating can also account for many of the observed electrostatic effects for ion channels. This is discussed for K^+ channels in Chapter 9.

1-1. The electron-gating model

Voltage-dependent ion channels play a central role in biological signaling and the mechanism for controlling ion conduction through the channel is of fundamental importance. In their classic 1952 publication, Hodgkin and Huxley suggested that a particle having a positive or negative charge, crossing the membrane, was associated with the transient changes in sodium conductance that occurred with changes in membrane voltage. Despite many important advances in understanding ion channels, the underlying mechanism of ion channel gating remains unknown. Gating is often described as a conformal change and has been associated with movement of the S4 voltage sensor. The transport of charge in the S4 region of the potassium channel protein was investigated using fluorescent labels by Mannuzzu et al., (1996). The results of this and later experiments have confirmed a charge movement, but the experiments have also been interpreted as a confirmation of rotation and sliding or other type of movement of the S4 voltage sensor and its positively charged arginine and lysine residues. However, an electron tunneling across these amino acids on the S4 could account for the observed fluorescence.

The model for ion channel gating, described in this book, is based on a stationary S4 voltage sensor with an electron tunneling across the arginine and lysine amino acid sites. Electron tunneling is a ubiquitous phenomenon at sub-nanometer distances and is known to occur in proteins over distances of less than 30 Å between donor and acceptor sites. Electron transfer rates have been measured, spanning 12 orders of magnitude for a 20 Å change in distance between donor and acceptor sites, and the presence of an intervening protein medium has been shown to substantially increase the rate of electron transfer (Moser et al., 1992; Page et al., 1999). Other studies have suggested that the structure of the intervening protein medium is critical in determining the rate of individual protein electron-transfer reactions (Beratan et al., 1992) and it is now clear that protein structures tune thermodynamic properties and electronic coupling to facilitate tunneling (Gray et al., 2003).

The electron-tunneling model has a selective tunneling path across the arginine and lysine amino acid residues of the S4 protein segment, known as the voltage sensor. The location of these residues had been deduced from cDNA and the amino acid sequence. The amino acid sequence of the sodium channel for the electric organ of the eel *Electrophorus electricus* was first determined by a group at Kyoto University, led by S. Numa (Noda et al., 1984). One of their findings was that arginine or lysine is occurring at every third residue in the S4 segment of transmembrane protein domains I through IV. In domains I through III, the S4 segment has from 4 to 5 arginine or lysine amino acids, spaced every third residue with a rise distance of 4.5 Å; in domain IV the S4 segment has 8 arginine or lysine residues, each having this spacing.

For the electron gating model, it is assumed that the transmembrane protein segments, including S4, have an α-helix geometry. To keep the model simple, it is further assumed that the S4 α-helix axis is perpendicular to the membrane, and that the electric field, due to membrane voltage, is parallel to the axis of the helix. The protein α-helix geometry was first described by Pauling et al., (1951). It has a rise per residue of 1.5 Å and a pitch height of 5.4 Å, corresponding to 3.6 residues per turn. The backbone radius is 2.3 Å. which does not include the outward sloping side chains. For the model, it is assumed that the sites for electron tunneling are the NH_3 groups at the end of the side chain. To obtain the tunneling distance, an estimated radius on the α-helix was determined for the tunneling centers. Using extrapolation of graphical data in a paper by Barlow and Thornton (1988), the radius to the NH_3

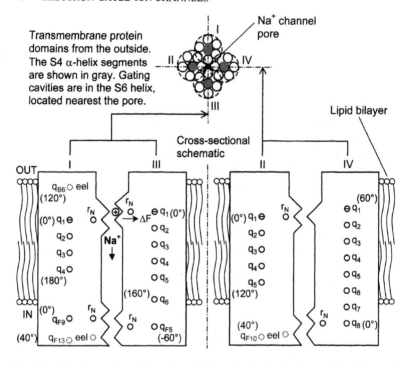

Fig. 1-1. The site map for sodium channel alpha subunit domains shows S4 arginine and lysine electron tunneling sites, designated by q_X, in relation to activation and inactivation energy barrier gates. An electron at a q_1 control site exerts a force ΔF on the sodium ion. This increases the energy required for the ion to pass the barrier, thus substantially lowering the transmission probability and resulting in a closed gate. Two additional energy barriers in sequence further reduce transmission probability and gate leakage when they are in the closed state. Far sites, designated q_{FX}, trap electrons (i.e. charge immobilization) and cause multiple long time constants for inactivation and gating current. The tunneling sites are for the giant axon of the squid *Loligo opalescens*. Additional sites for *Electrophorus electricus* (eel) are shown by dotted circles. The back site q_{B6} (for eel) with a time constant of tens to hundreds of milliseconds may alter the gating kinetics. To simplify mapping, the sites are shown along the S4 α-helix axis and have spacing proportional to the rise distance between the sites. The cross-sectional angle ϕ (Table 8-1) for the tunneling site on the α-helix is shown in parenthesis with $0°$ closest to the channel axis. The distance across the protein is ~45 Å, based on the total rise distance for the α-helix to cross the site map. See Chapter 8 for further discussions.

nitrogen center for arginine was estimated to be about 4 Å. Using a radius of 3.97 Å and other data given above, the distance r between tunneling centers, at every third residue, was calculated to be 6.0 Å at an angle θ of 41.4 degrees with the electric field. The second decimal place radius adjustment makes the calculated tunneling distance an even 6.0 Å.

1-2. Electron gating of a sodium channel

To introduce the operation of the electron-gating model, it is convenient to represent ion channel gating as a series of switches. An overview of the sodium channel pulse generating process is illustrated by the circuit and timing diagram of Fig. 1-2. Ion channel gates are represented by series switches, all of which must be closed for current to flow. A closed switch represents an energy barrier in the low-energy open state. Associated with each switch is an electron tunnel track having N tunneling sites. A switch is opened by the presence of an electron at a control site at the end of a tunnel track. When an electron is at the control site, it causes an attractive, lateral, Coulomb force on the positively charged ion at an energy-barrier gating cavity in the channel. This force, in conjunction with the geometry at the gating site, increases the energy barrier height, thereby blocking ion transport.

In the model the sodium channel has three activation energy barriers, each with an activation control site and a corresponding electron tunnel track (only one is shown) with multiple tunneling sites. An inactivation energy barrier and its control site are located near the inside edge of the membrane protein. A tunnel track represents the path of an electron crossing the arginine or lysine amino acids on a S4 segment of the α-helix protein. The tunneling path is diagonal across the α-helix at an angle θ with the electric field, but for simplified illustration, it is shown along the S4 α-helix axis. A site map for sodium channel gating, based on amplified electron tunneling, is shown in Fig. 1-1.

1-3. Timing

The timing diagram of Fig. 1-2 shows the sequence of events. With a large negative or positive patch clamp voltage, most of the electron probability is at the end sites and the switches can be treated as being either open or closed. At time t_0-, the membrane is in steady state at the resting potential with all activation switches (m) open and the inactivation switch (h) closed. At time t_0, a depolarizing patch clamp voltage pulse is applied across the membrane. This causes the tunnel track electrons to be attracted away from their control site positions and activation switches to close. The time delay from t_0 to t_1 is for the last electron to leave an activation control site, thereby allowing current flow.

A time delay of t_{1+} to t_2 is required for an electron to tunnel across seven sites to the inactivation control site and open the inactivation switch (h) at time t_2. During this interval, all switches are in the closed state and ion current flows. When the patch clamp voltage is returned to

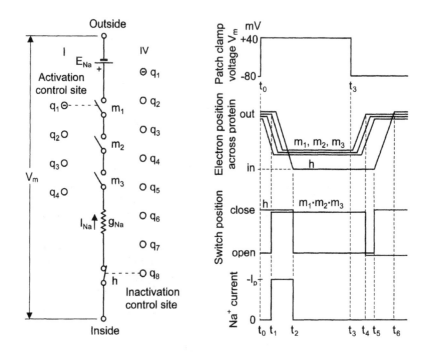

Fig. 1-2. Circuit and timing diagram illustrating sodium channel operation. Arginine or lysine electron tunneling sites on the S4 transmembrane protein segment are designated by q_1 through q_4 for domain I, and by q_1 through q_8 for domain IV. The sites for domains II and III are not shown. Switch positions are for a large negative resting potential with an electron at each of the q_1 sites. When the membrane is depolarized the force on each electron acts to cause tunneling across the tunnel track sites. Switches m_1, m_2, and m_3 close after the electron leaves each activation control site. Sodium ions flow inward until the electron in track IV reaches the inactivation control site and opens switch h. The timing diagram shows the switching sequence for a positive patch clamp voltage step, followed by a return to resting potential. The sodium ion current pulse is activated after an initial delay for all electrons to leave the activation control sites. For the large voltage change shown, the maximum width of the current pulse is determined by the time it takes the electron in track IV to tunnel across seven sites to the inactivation control site. The width of the current pulse may be less than the inactivation delay due to the sodium ion driving potential ($V_m - E_{Na}$) reaching a near zero value before the electron arrives at the q_8 inactivation control site.

the resting potential at time t_3, the electrons are repelled towards the q_1 control sites. There is a time delay of t_3 to t_4 for the first electron to cross the $N - 1$ sites to the activation control site. Upon its arrival, an activation switch is opened, blocking any possibility of current flow. Because of repulsive forces, return of the inactivation electron to the resting site is delayed until all of the activation electrons return to their control sites. The repulsive forces between electrons in adjacent tracts affect timing by delaying late-starters at the end sites. This results in a probable serial departure of electrons from the control sites and a probable delayed departure of the inactivation electron. The sodium channel is not completely ready for retriggering until the return of the inactivation electron to its q_1 site at time t_6. In normal operation, the recovery time from t_3 to t_6 is controlled by the magnitude of current flowing in the delayed rectifier potassium channels that discharge the membrane capacitance.

1-4. Sodium channel current

Each sodium channel current pulse is composed of a large number of ions. The number of ions in a pulse depends on the membrane voltage and the gate open time. It is assumed that the ions transit the gates in single file and only one ion can be in a gating cavity at a time. As a result, the requirement of the circuit diagram, that all switches must be closed (all gates must be open) simultaneously for current to flow, should be (as a first approximation) a valid representation. If the operation of each switch is statistically independent of all the others, then the probability of all switches being closed at a given time is the product of the probability for each switch being closed. Describing this in terms of three identical activation tunnel tracks and an inactivation track, gives the equation

$$P_o = \left(1 - P_1\right)^3 \left(1 - P_8\right), \tag{1.1}$$

where P_o is the probability that all switches are closed (channel is open) and P_1 and P_8 is the probability of tunnel track electrons being at the indicated control sites. Using this equation, the average current for the sodium channel is given by

$$I_{Na} = g_{Na}\left(1 - P_1\right)^3 \left(1 - P_8\right)\left(V_m - E_{Na}\right), \tag{1.2}$$

where g_{Na} is the maximum conductance of the channel. The voltage polarity is for the customary negative sodium channel current. It is assumed that the tunnel-track electrons function sufficiently independent of each other for gating to be treated as the product of the probabilities and that any changes in timing and midpoint voltage, due to Coulomb interaction of the electrons, can be treated separately.

1-5. Sensitivity

One of the challenges in developing the model was accounting for the sodium ion channel voltage sensitivity. Early in the investigation, it was realized that it would not be possible to have an electron gating model without discovering a mechanism for amplification. Initially it was assumed that the membrane voltage establishes an electric field over a distance x_p of about 45 Å. This distance was estimated from the amino acid sequence for the sodium channel alpha subunit using a site map (Fig. 1-1) that includes far sites for electron tunneling to arginine and lysine. By including these far sites of the S4 transmembrane protein segment in the site map, an explanation could also be provided for the long time constants of gating current and inactivation. The distance x_p is one of the calibration variables in the model. It is the effective distance for determining the electric field across the protein and is only roughly equivalent to the geometric distance across the protein.

The rise distance x_r between arginine or lysine sites, at every third residue, is 4.5 Å for the α-helix geometry. From this and the estimated 3.97 Å radius, a center-to-center tunneling distance of 6.0 Å was calculated. The method for calculating this distance and the tunneling distances between far sites is given in the appendix. The ratio of $x_r/x_p = \eta$ acts as a voltage divider, so that only a small fraction of the membrane voltage is across adjacent arginine tunneling site centers. For the data shown in Table 1-1, the effective distance across the protein is: $x_p = 42.5$ Å and the ratio x_r/x_p is 0.106. With this attenuation of membrane voltage, an amplification of about 25 is needed to have agreement with the sodium ion channel sensitivity, first determined by Hodgkin and Huxley (1952) for the squid giant axon.

1-6. Amplification and negative conductance

A well-known, fundamentally important characteristic of sodium channels is negative conductance. This characteristic is illustrated by the I-V curve (Fig. 1-3). In the model, electron-gated energy barriers have a steeply increasing open-state probability with membrane depolarization. As the membrane is depolarized, the ionic driving potential ($V_m - E_{Na}$) decreases. The steep positive slope of the open-gate probability curve, combined with the decreasing driving potential, produces a negative slope for the sodium ion current. This is the negative conductance region. The curves were plotted from Eqs. 7.4 and 7.5 and the electron gating model equations of Table 6-1C with amplification of $h_w = 25.2$. This amplification gave agreement with the Hodgkin and Huxley rate constant β_m.

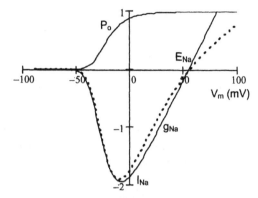

Fig. 1-3. Illustration of well-known I-V curves for the sodium channel, plotted with equations for the electron-gating model. The sodium ion current increases with increasing open-channel probability P_0 until saturation, and then the current decreases with a slope determined by the maximum channel conductance. The solid line is for a linear model with a slope g_{Na} for the fully open channel. The dotted line is for a gated-flux model, using the GHK current equation. The steeply increasing current with a decreasing ionic driving potential (negative conductance) allows triggering of an action potential near the start of negative conductance.

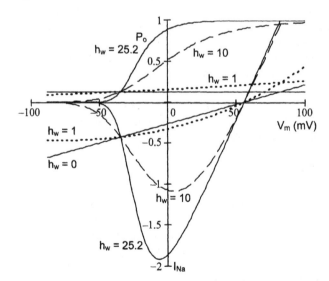

Fig. 1-4. Open-channel probability and sodium ion current are shown for four values of amplification. The negative conductance disappears when there is no amplification ($h_w = 1$); however, gating still responds to membrane potential. When h_w equals zero, there is no gating. The slope of the curve is then determined by the fraction of permanently open channels. Here the open-channel probability P_0 (with $h_w = 25.2$) for activation is equal to m^3 in the Hodgkin-Huxley model.

Negative conductance requires amplification

It is interesting to see what happens when amplification in the electron-gating model is reduced. When the amplification is reduced to unity, there is no longer a negative conductance and the sodium channels cannot trigger a current pulse. This is what would be expected for an amplifying device. An amplifier with a positive feedback loop can be arranged to show negative conductance at its output. Without amplification, negative conductance cannot be obtained and generating a sustained oscillation is not possible. Amplified positive feedback drives an amplifier into saturation. The sodium channel has an amplified positive feedback loop that drives it into saturation; i.e. membrane depolarization opens the activation gates, allowing more sodium current to flow, which further increases the depolarization until there is no longer a driving force ($V_m = E_{Na}$). Then, the electrons, in domain IV tunnel-tracks, arrive at the inactivation control sites, creating energy barriers that close the channels. This ends the Na^+ current pulse that charges the membrane capacitance.

The pulse generating cycle is completed by the delayed rectifier potassium channels, which discharge the membrane capacitance and return membrane voltage to the resting potential. In the discharging process there is an overshoot to the potassium equilibrium potential E_K, followed by a recovery time of a few milliseconds to the resting potential. The minimum recovery time is controlled by the potassium channel time constant, which is near its maximum at the resting potential. An analysis of potassium channels in terms of the electron-gating model is given in Chapters 7 and 9.

The amplification factor referred to above as h_w is for a tunneling voltage v. The membrane voltage V_m is multiplied by attenuation factor η to give the tunneling voltage.

1-7. Model parameters

Certain constants and parameters are used throughout the model. These are listed in Table 1-1. As determined by microwave spectroscopy (Part II), the arginine side chain has two NH_3 groups, each with a different inversion frequency and a different amplification for electron tunneling. The lower inversion frequency, corresponding to a higher amplification, gave a greater amplifying range for the rate constants before saturation and a better match with the Hodgkin-Huxley data. This frequency was used for the arginine calculations and the associated parameter values are listed in Table 1-1. Parameter values for the higher inversion frequency are for lysine. They are listed in Table 3-1.

TABLE 1-1 Parameter values and symbols

Physical Constants

Symbol	Value	Description
h	6.6262×10^{-34} J s	Planck constant
	4.1357×10^{-15} eV s	Planck constant (for energy in eV)
e	1.6022×10^{-19} C	Elementary charge
m_e	9.1095×10^{-31} kg	Electron rest mass
N_A	6.0221×10^{23} mol^{-1}	Avogadro constant
R	8.3145 $J mol^{-1} {}^{\circ}K^{-1}$	Gas constant
ε_0	8.8542×10^{-12} $C V^{-1} m^{-1}$	Permittivity of vacuum
c	2.9979×10^{8} ms^{-1}	Velocity of light in vacuum
k	1.3807×10^{-23} $J {}^{\circ}K^{-1}$	Boltzmann constant
	8.6175×10^{-5} $eV {}^{\circ}K^{-1}$	Boltzmann constant (for energy in eV)
amu	1.6606×10^{-27} kg	Atomic mass unit

The following values are used throughout the book unless otherwise noted.

Symbol	Value	Description
x_r	4.5 Å	Rise distance for every 3rd residue on α-helix
x_p	42.5 Å	Electric field distance for Na^+ channel protein
r	6.0 Å	Distance between tunneling site centers
θ	41.4 °	Tunneling angle with electric field, $r = 6$ Å
r_0	4 Å	Width of electron energy well
r_d	0.27 Å	Electron energy well displacement distance
η	0.106	Membrane voltage attenuation for Na^+ channel
U	5.75 eV	Electron energy well depth
ΔE_0	5.93×10^{-2} meV	Energy for the 14.3 GHz inversion frequency
ΔE_1	2.70 meV	Energy difference for the first excited state
h_w	25.2	Amplification factor at 6°C
a_r	4.74	Energy ratio for amplification window, $h_w = 25.2$
E_{dmax}	1.32 meV	Maximum displacement energy
$k'T$	24 mV (kT/e_0)	Thermal energy per $1e$ charge at 6°C
β_0	1.025 $Å^{-1} eV^{-1/2}$	Electron tunneling constant
k_f	8.14×10^{-4} $eV Å^{-2}$	Force constant for molecular spring model
K_0	1000 s^{-1} (also 1 ms^{-1})	Rate constant factor at 6°C
k'	8.6175×10^{-2} $mV {}^{\circ}K^{-1}$	Boltzmann constant per $1e$ charge
R	1.6×10^{13} ohms	Tunneling resistance between sites at 6°C
C_0	168×10^{-18} F	Capacitance factor
e_0	$1e$	Electron charge for energy in eV
G_{oh}	50 meV	Sodium channel open-gate energy barrier
ΔG_h	180 meV	Sodium channel energy barrier change
K_{max}	10^{13} s^{-1}	Maximum ET rate for $\Delta G^{\circ} = \lambda$ at contact distance
λ	1.431 eV	Reorganization energy for electron tunneling
σ	2.77	Calibration for time constant capacitance
T	$6 + 273°K$	Temperature used in calculations

TABLE 1-1A Additional symbols

Symbol	Description
V_m	Membrane voltage
E_r	Membrane resting potential
V	Displacement voltage as defined by Hodgkin and Huxley
v	Tunneling voltage - potential between tunneling site centers
i_d	Displacement current through tunneling resistance R
I_{Na}	Average current through sodium ion channel
E_R	Energy dissipated in tunneling resistance R
H	Reciprocal sensitivity factor for membrane voltage
H'	Reciprocal sensitivity factor for tunneling voltage
C_d	Displacement capacitance
C_{dp}	Peak displacement capacitance
C_τ	Time constant capacitance
$C_{\tau p}$	Peak time constant capacitance
α_e	Electron-tunneling rate constant – towards control site
β_e	Electron-tunneling rate constant – away from control site
γ_α	Gating energy barrier distortion for α rate constant
γ_β	Gating energy barrier distortion for β rate constant
c_i	Curve correction coefficients for α rate constants
q_n	A *near site* located on S4 or a charge located at tunneling site n
q_{FX}	A *far site* located X residues from last near site towards cytoplasm
q_{BX}	A *back site* located X residues from the control site
r_N	Location of dipoles producing open-gate energy barriers
N	Number of tunneling sites in a tunnel track
f_0	NH_3 inversion frequency
f_1	First excited vibrational state inversion frequency
R_0	Contact resistance in SETCAP model
β	Distance-decay constant
τ_e	Time constant for electron tunneling
τ_{ep}	Peak time constant for electron tunneling
ϕ	Cross-sectional angle of tunneling site on α-helix referenced to q_1
ΔG^o	Electron free energy
λ_e	Wavelength for electron in energy well ground state
Q'_{10}	Temperature factor for ion channel pore with a 10°C rise
Q_{10}	Total temperature factor for rate constants with a 10°C rise
K_{ET}	Electron-transfer rate as determined by free energy $-\Delta G^o$
a'	Forward coupling factor for sodium inactivation gating
P_o	Open channel probability
P_1	Probability of a tunnel track electron being at activation control site
P_{ob}	Probability of a modulated energy barrier being in the open state
γ_o	Open-gate distortion factor
γ_c	Closed-gate distortion factor due to gate leakage

Chapter 2

DEVELOPING A MODEL

2-1. A single electron two-site model

In developing an electron-tunneling model, equations were first derived for a tunnel track having two identical tunneling sites and a single tunneling electron. This was then expanded in Chapter 3 to include N tunneling sites. In deriving the equations, certain assumptions and simplifications were made. It was assumed that the only electric field crossing the tunneling sites was due to membrane voltage and that the electric flux was parallel to the axis of the α-helix. Square energy wells could be used instead of parabolas, because the energy change due to membrane voltage is very small ($< 2kT$), thus allowing a square well model for the ground state. The following derivations are for steady state.

Starting with the Time Independent Schrödinger Equation for an electron in a square energy well with finite sides

$$\frac{d^2\psi}{dx^2} - \frac{8\pi^2 m_e\left(V' - E\right)}{h^2}\,\psi = 0, \tag{2.1}$$

where E is kinetic energy of the electron and V' is the potential energy of the electron in the well. In the energy barrier region of $V' > E$, a general solution is

$$\psi = D \exp\left(-\frac{2\pi}{h} \sqrt{2m_e \left(V' - E \right)}\, x \right). \tag{2.2}$$

For two energy wells, with an energy barrier of width x between them, the transmission probability for the barrier is given by the ratio of $|\psi|^2$ evaluated on the acceptor side of the barrier, to $|\psi|^2$ evaluated on the donor side of the barrier.

$$P_T(x) = \frac{|\psi|^2_{x_1}}{|\psi|^2_{x_0}} = \exp\left(-\frac{4\pi}{h} \sqrt{2m_e(V'-E)}\, (x_1 - x_0) \right) \tag{2.3}$$

Rewriting Eq. 2.3 in a simplified form gives the transmission probability for a distance x.

$$P_T(x) = \exp\left(-\beta_0 \sqrt{U}\, x \right) \tag{2.4}$$

In the equation, the term U replaces $V - E$ and represents the energy well depth in electron volts. The total energy of the donor electron in the energy well is $-U$, with electron energy having a negative value. The term β_0 replaces the Schrödinger equation constants and includes a factor of 10^{-10} to allow distance to be expressed in angstroms. The elementary charge converts energy units to electron volts. The term β_0 is referred to here as the electron-tunneling constant.

$$\beta_0 = \frac{4\pi}{h} \sqrt{2m_e e}\, 10^{-10} \tag{2.5}$$

It has a calculated value of $\beta_0 = 1.025$ Å^{-1}eV$^{-1/2}$. The width of the energy well depends upon the well depth U. It can be determined approximately from the well-known (particle-in-a-box) equation for the wavelength of an electron in the ground state of a square energy well.

$$\lambda_e = \frac{h}{\sqrt{2m_e e U}\, 10^{-10}} \tag{2.6}$$

The well width is given by $\lambda_e/2$. Combining the previous two equations gives the electron wavelength in terms of the electron-tunneling constant and energy

$$\lambda_e = \frac{4\pi}{\beta_0 \sqrt{U}}. \tag{2.7}$$

At low free energies, the sensitivity of electron tunneling to a potential v across the tunneling sites can be described by exponential rate constants. In the electron-gating model, the voltage sensitivity is represented by the rate constants $\alpha_e(v)$ and $\beta_e(v)$. Rate constant $\beta_e(v)$ is for the electron-tunneling rate away from the gating control site and rate constant $\alpha_e(v)$ is for the tunneling rate towards the gating control site.

$$q_1 \underset{\alpha_e}{\overset{\beta_e}{\rightleftharpoons}} q_2$$
$$O \qquad O$$

This polarity allows the rate constants for electron tunneling to correspond to the Hodgkin-Huxley rate constants (i.e. $\beta_e \rightarrow \beta_m$). The equations for rate constants are developed in Chapter 3.

For a charge of one electron, tunneling across an energy barrier with a potential v between the two sites, the rate constants can be expressed as: $\alpha(v) = K\exp(e_0 v/2kT)$ and $\beta(v) = K\exp(-e_0 v/2kT)$. Incorporating Eq. 2.4 as part of factor K, then replacing v with ηV_m gives the α rate constant as

$$\alpha_e(V_m) = K_a \exp\left(\frac{e_0 \eta V_m}{2kT}\right) \exp\left(-\beta_0 \sqrt{U} x\right). \tag{2.8}$$

The factor η scales membrane voltage V_m to the potential v between the tunneling site centers. K_a is a rate-constant calibration factor. The rate constant decreases exponentially with increasing edge-to-edge tunneling distance x, as indicated by the *distance factor*. For the equations, it is assumed that the free energy of the donor electron is less than $2kT$.

2-2. Amplification

A model for amplification was developed based upon membrane voltage modulation of the energy barrier width, using a simple harmonic oscillator model for molecular stretching. The energy barrier width was defined as

$$x = r - r_0 - 2\delta, \tag{2.9}$$

where x is the energy barrier width (or edge-to-edge tunneling distance),

r is the center-to-center distance between tunneling sites, r_0 is the width of an energy well, and δ is a displacement distance due to molecular stretching. A simple harmonic oscillator model for molecular stretching uses Hooke's Law: $F = k_f \delta$. For this amplification model, F is the force due to the electric field acting on a donor electron and k_f is a force constant. When the electron is at the q_1 tunneling site, a positive membrane voltage causes an attractive force on the electron towards site q_2. This results in stretching of the molecule (as a spring) and a displacement δ, thus reducing the energy barrier width. The force on a charge in an electric field is given by $F = qE$. The average force on a donor electron (with a 0.5 probability of being at the site) in the direction of tunneling is given by the equation

$$F = \frac{-e_0 V_m \cos\theta}{2x_p}.$$

(2.10)

This is equated to the restoring force in Hooke's Law of $-k_f\delta$. Solving for δ and substituting the terms into the expression for the energy barrier width (Eq. 2.9) gives

$$x = r - r_0 - 2\left(\frac{e_0 V_m \cos\theta}{2k_f x_p}\right).$$

(2.11)

The term x_p is the effective distance across the channel protein, which determines the magnitude of the electric field, and θ is the angle between the electric field and the direction of electron tunneling. Substituting the terms for Eq. 2.11 into Eq. 2.8 gives

$$\alpha_e(V_m) = K_a \exp\left(\frac{e_0 \eta V_m}{2kT}\right) \exp\left[-\beta_0 \sqrt{U}\left(r - r_0 - \frac{e_0 V_m \cos\theta}{k_f x_p}\right)\right].$$

(2.12)

The attenuation factor for membrane voltage is given by

$$\eta = \frac{x_r}{x_p} = \frac{r\cos\theta}{x_p}.$$

(2.13)

The potential v between the tunneling centers is equal to the attenuation factor η times the membrane voltage V_m. Substituting the terms for η

into Eq. 2.12, simplifying, and rearranging terms, gives the more useful expression

$$\alpha_e(V_m) = K_a \exp\left[-\beta_0\sqrt{U}\,(r-r_0)\right]\exp\left[\left(1+\frac{2\beta_0\sqrt{U}kT}{k_f r}\right)\frac{r\cos\theta}{x_p}\frac{e_0 V_m}{2kT}\right]. \quad (2.14)$$

From this equation, an amplification factor h_w is defined as

$$h_w = 1 + \frac{2\beta_0\sqrt{U}kT}{k_f r}. \quad (2.15)$$

When the force constant is large ($k_f > 1$), the amplification factor is unity. Incorporating h_w and η into Eq. 2.14 gives the simplified expression

$$\alpha_e(V_m) = K_a \exp\left[-\beta_0\sqrt{U}\,(r-r_0)\right]\exp\left(\frac{h_w e_0 \eta V_m}{2kT}\right). \quad (2.16)$$

Because membrane voltage is in mV, it is convenient to define a term k' that is equal to the Boltzman constant per electron with units of mV per °K. This gives: $k' = 8.6175 \times 10^{-2}$ mV °K^{-1} and $k'T = 24$ mV @ 6°C. The symbol k' is used to reduce the number of terms in the expressions.

From Eq. 2.16, the β_e rate constant for electron tunneling can be written as

$$\beta_e(V_m) = K_a \exp\left[-\beta_0\sqrt{U}\,(r-r_0)\right]\exp\left(-\frac{h_w \eta V_m}{2k'T}\right). \quad (2.17)$$

An important finding described in Section 5-4 is that, for sodium channel activation gates, there is negligible ion channel distortion of the electron-tunneling rate constant β_e. Therefore, the voltage sensitivity for rate constant β_e should be the same as the experimentally determined sensitivity for β_m. From this, the amplification can be determined by equating the sensitivity for β_e with the corresponding 1/18 voltage sensitivity for the Hodgkin-Huxley rate constant β_m.

$$\frac{h_w \eta}{2k'T} = \frac{1}{18} \quad (2.18)$$

Solving for the amplification using $\eta = 0.106$ (Eq. 2.13) gives

$$h_w = \frac{1}{18}\left(\frac{2k'T}{\eta}\right) = 25.2 . \tag{2.19}$$

The sensitivity for β_m was used for matching, because (according to the model) the α_m sensitivity is reduced, due to the electron probability being spread over four or five arginine or lysine tunneling sites. The above-calculated amplification is for a temperature of 6°C. This temperature is used throughout the model to facilitate matching equations with the experimental data of Hodgkin and Huxley.

2-3. A small force constant

An unknown factor determining the amplification h_w, in Eq. 2.15, is the force constant. Solving the equation for the force constant gives

$$k_f = \frac{2\beta_0 \sqrt{U} kT}{(h_w - 1) r} = 8.14 \times 10^{-4} \, eV\text{Å}^{-2} . \tag{2.20}$$

The amplification process was first described as due to molecular stretching. In one analogy, a molecular spring at the q_1 tunneling site has no tension with $V_m = 0$. With a positive voltage change in V_m and a donor electron at q_1, the spring first stretches, reducing the tunneling distance, then it returns to zero tension after the donor electron tunnels to the q_2 site. The above-calculated force constant, in SI units, is $1.3 \times 10^{-2} \, Nm^{-1}$. Polar molecules typically have force constants several orders of magnitude larger than this, with associated absorbance spectra in the infrared region and very small stretching distances. What could cause this very small force constant and a large (~ 0.3 Å) stretching or compression distance?

To find an answer to this question, a number of calculations were made. The first calculation was to determine the maximum stretching or compression distance of the spring. This corresponds to the maximum displacement energy from the unstretched $v = 0$ reference. From Eq. 2.11, the displacement distance δ, for an electron at a donor tunneling site, is given by

$$\delta = \frac{e_0 V_m \cos\theta}{2k_f x_p} . \tag{2.21}$$

Rewriting this equation in terms of the tunneling voltage by replacing the terms for the electric field in the direction of tunneling with v/r gives

$$\delta = \frac{e_0 v}{2 k_f r}.$$
(2.22)

The term $e_0 v / 2r$ represents the average force on an electron, having a 0.5 probability, which is stretching the spring a distance δ. In the two-site model, the electron has a 0.5 probability of being at the donor site with $v = 0$. The term $e_0 v$ is the energy absorbed from the electric field for transferring an average charge of $e_0/2$ through a potential v across distance r. There is, however, an upper limit on the energy absorbed from the electric field and a corresponding maximum displacement distance r_d. This is because the probability of the electron being at the q_1 site decreases as the potential v across the tunneling sites becomes more positive. The force acting to stretch the spring to the maximum displacement r_d is equal to the force on the donor electron, times the probability of the electron being present at the donor tunneling site.

$$k_f r_d = \frac{P_1 e_0 v}{r} = \frac{E_{dmax}}{r}$$
(2.23)

After rearranging terms,

$$r_d = \frac{E_{dmax}}{k_f r} = 0.27 \, \text{Å}.$$
(2.24)

The term r_d is used to represent the maximum displacement of the spring, instead of δ_{max}. The displacement energy E_{dmax} is the energy required to displace all of the ensemble average charge from the $v = 0$ reference. The total displacement energy required to transfer a charge of one electron from a donor site to an acceptor site is $2E_{dmax}$ (for a large $-v$ to $+v$). The maximum displacement energy can be calculated with Eq. 3.24. The change in the displacement energy with tunneling voltage is illustrated by the graph in Fig. 3-2.

2-4. Calculating frequencies

Here we are interested in calculating the energies at the donor tunneling site. A spring with a force constant k_f and a maximum displacement r_d

undergoes a potential change of $v = 0$ to a saturating value of $+v$. This corresponds to stretching of a molecular spring by r_d and an energy of $\Delta E_0/2$.

$$\frac{\Delta E_0}{2} = \frac{1}{2} k_f \left(r_d\right)^2$$

$$\Delta E_0 = 5.93 \times 10^{-5} \, eV \tag{2.25}$$

A potential change of $v = 0$ to a saturating $-v$ corresponds to a compression of a spring by $-r_d$, and an additional energy $\Delta E_0/2$. For the total energy ΔE_0, a resonant frequency was calculated using the equation $\Delta E = hf$.

$$f_0 = \frac{\Delta E_0}{h} = 14.3 \, GHz \tag{2.26}$$

The energy change required for transferring a charge of one electron to the adjacent tunneling site is equal to twice the maximum displacement energy or $2E_{dmax}$. Using the equation $\Delta E = hf$ again, the frequency can be calculated for this transition. A small correction with a transition factor n_t (Eq. 2.29) is incorporated, so that the frequency calculation corresponds to Fig. 2-1.

$$f_1 = \frac{2E_{dmax}}{n_t h} = 650 \, GHz \tag{2.27}$$

It is clear from the above calculations, that amplification based on the molecular stretching model, has an associated microwave resonance in the region of 12–18 GHz. The tunneling sites for the model are the NH_3 groups at the end of the arginine and lysine side chains. A resonance in this microwave frequency range might be due to a hindered motion with tunneling through an energy barrier, a hindered torsional motion with tunneling through a barrier, or to molecular rotation. However, the rotational frequencies of NH_3 are reported to be considerably higher than the inversion frequency of the ground state.

The above-calculated frequencies are remarkably close to the umbrella inversion resonance of gas-phase NH_3 at 23.786 GHz (0.793 cm^{-1}) and the first excited vibrational resonance at 1080 GHz (36 cm^{-1}). This suggested that the inversion resonance was continuing to occur with NH_3 attached to a carbon atom at the end of arginine and lysine amino acid side chains. The original calculation for f_0 was 21 GHz and the amplification h_w, distance x_p, and energy were somewhat different than shown above. The values have been adjusted to provide agreement with the experimentally determined frequency of $f_0 = 14.3$ GHz and the scaled down frequency f_1. The frequency f_1 was calculated by scaling down the 1080 GHz frequency for NH_3 in the gas-phase with the same scaling factor determined for the inversion frequency f_0. Experimental determination of the frequencies is described in Part II of the book.

2-5. Amplification by NH_3 inversion resonance

A model based upon NH_3 inversion resonance was developed to replace the force constant and to account for the electron tunneling amplification using the NH_3 inversion frequency. In this model, membrane voltage modulation of the energy barrier width is not due to conventional molecular stretching; instead, the process involves the NH_3 umbrella inversion.

Incorporating the equation $\Delta E_1 = hf_1$ into Eq. 2.27 gives

$$\Delta E_1 = \frac{2E_{dmax}}{n_t}.$$ (2.28)

where the energy transition factor n_t is

$$n_t = 1 - \frac{\Delta E_0}{\Delta E_1}.$$ (2.29)

Equation 2.15 was modified by replacing the force constant with parameters associated with the NH_3 inversion. It follows that:

rearranging Eq. 2.25 gives

$$k_f = \frac{\Delta E_0}{\left(r_d\right)^2}.$$ (2.30)

Substituting Eq. 2.30 into Eq. 2.15 gives an equation for amplification in terms of the energy ΔE_0.

$$h_w = 1 + \frac{2\beta_0\sqrt{U}\,r_d^2\,kT}{\Delta E_0\,r}$$ (2.31)

In terms of the ground-state inversion frequency, the amplification is

$$h_w = 1 + \frac{2\beta_0\sqrt{U}\,r_d^2\,kT}{hf_0 r}.$$ (2.32)

Amplification theory

From Eq. 2.31, it is evident that ΔE_0 and r_d are principal factors in the NH_3 amplification of electron tunneling. An investigation was made to understand how the amplification could occur, based on the physics of the NH_3 inversion. Many of the detailed aspects of NH_3 behavior in this environment are unknown, but the following overall description is inferred from equations for the model, the observed frequencies, and the well-studied behavior for NH_3 in the gas phase.

In NH_3 inversion, the energy well, for the molecule, is divided into two parts with an energy barrier between them (Fig. 2-1). Mixing of symmetric and antisymmetric wave functions causes the well-known splitting of the ground state and the higher energy vibrational states. When the wave functions are symmetric, a lower energy level E_{0+} results and this also corresponds to a lower energy level E_{1+} in the first excited vibrational state. Antisymmetric wave functions correspond to the higher energy levels E_{0-} and E_{1-}. The electrons bonding the three hydrogen nuclei of NH_3 to the nitrogen tend to be slightly closer to the nitrogen than to the hydrogen nuclei, making the hydrogen atoms slightly positive. When a donor electron (D) is present at a tunneling site, it orbits the nitrogen atom and would likely spend much of the time in the lower potential energy region between the three hydrogen atoms. This region is reported to act like a cavity for an excess electron. In solutions of metals in liquid ammonia, the ammoniated electron $e_{(am)}$ appears to consist of an electron trapped in a 0.3 nm cavity of NH_3 (Sisler, 2002). For the model, it is assumed that this region, between the three hydrogen atoms, acts like a cavity for the tunneling electron. When a donor electron is in the cavity, it undergoes the vibrational motions and energies imparted by NH_3 inversion. The important vibrational energies for amplification are shown in Fig. 2-1.

Interactions with an electric field

There are several types of interactions with an electric field. One interaction is for a change in the potential energy of the donor electron due to a displacement of the electron orbit (and the electron energy well) and caused

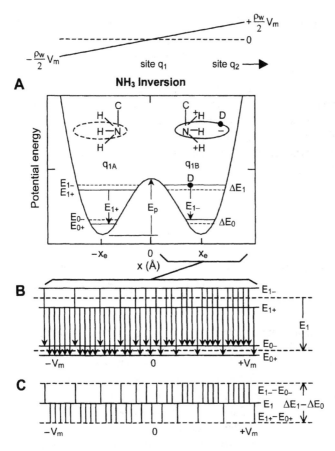

Fig. 2-1. The split energy well, shown in A represents inversion of a NH_3 group at the end of an arginine or lysine side chain. A donor electron (D) orbits the nitrogen atom. As the NH_3 inverts, at a frequency f_0, the donor electron shifts its orbit as schematically shown in A. Amplification occurs by altering the probable decay path, from the first excited vibrational state, with a change in membrane voltage. With membrane voltage V_m equal to zero, there is an assumed equal probability for the donor electron to decay from E_{1-} or E_{1+} as shown for energy well q_{1B} in B and C. If V_m goes positive, the probability increases for the electron to be in well q_{1B} and to decay from E_{1-} to ground state E_{0-}. When NH_3 is in the higher energy first excited state, the transition is to the higher energy ground state and when in the lower energy state to the lower energy ground state. The energy difference in the two paths is $\Delta E_1 - \Delta E_0$. This gives the maximum energy change a tunneling electron can receive with a large negative to positive change in the membrane voltage. NH_3 inversion imparts an oscillating displacement of $2r_d$ to the electron energy well, as illustrated in Fig. 4-1. The energy ΔE_0 and displacement r_d are factors determining the amplification (Eq. 2.31). The energy change for the tunneling electron of $\Delta E_1 - \Delta E_0$ determines the bandwidth.

by inversion of the NH_3 molecule. The electric field is represented by ξ.

$$\Delta E_e = 2r_d e \xi \tag{2.33}$$

A second interaction with an electric field is due to the electric dipole moment of the NH_3 molecule. The ammonia molecule has an unusually high polarizability and high sensitivity to an electric field because of the small value (ΔE_0) of the ground-state energy splitting (Feynman, 1965). In an external electric field, the dipole moment causes an energy increase given by $\Delta E_0'$.

$$\Delta E_0' = \Delta E_0 \sqrt{1 + \left(\frac{2\mu\xi}{\Delta E_0}\right)^2} \tag{2.34}$$

The dipole moment for NH_3 (gas phase) is $\mu = 1.47$ debyes (4.9×10^{-30} Cm) and from Eq. 2.25, ΔE_0 is 5.93×10^{-5} eV. For a membrane voltage of 20 mV, the electric field is about $\xi = 5 \times 10^6$ Vm^{-1}.

With a membrane depolarization, as in Fig. 2-1, the electric field acts to reduce the distance between the plane of the hydrogen atoms and the nitrogen atom, thereby increasing the frequency and energy of vibrations in well q_{1B}. The field, however, causes displacement of the electron orbit in the opposite direction and this would act to increase the distance between the plane of the hydrogen atoms and the nitrogen atom, thus canceling much of the frequency increase. The strong interaction between the donor electron and the dipole moment of NH_3 reduces the effect of the electric field on the dipole moment.

With an external electric field as in Fig. 2-1, the force acting on the donor electron is aligned opposite to the direction of the field and the force acting on the dipole is in the direction of the field. The electron and the NH_3 dipoles are tightly bound, but there would likely be a small change in the displacement between them in the electric field, which could result in a change in potential energy for the donor electron.

Fig. 2-2. A donor electron in the ground state of an arginine NH_3 group is thermally excited to a higher energy state with a short lifetime. The electron then decays to the metastable first-excited vibrational state. When it decays back to the ground state, the electron may tunnel to an adjacent acceptor site carrying with it the energy E_{1-}. Excitation and decay may also occur by an alternate path, giving the tunneling electron energy E_{1+}. The probable decay path is controlled by the electric field acting on the donor electron and the NH_3 dipole moment. The electric field determines the vibrational mode of the ground state and also the higher energy states and the energy for a tunneling electron.

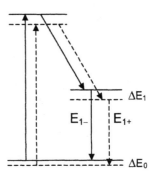

Mode switching and amplification

A well-known property of NH_3 is that it can vibrate in either of two stationary states. These stationary states are referred to here as oscillation modes. In the ground state the modes are separated by energy ΔE_0. When a membrane electric field crosses an NH_3 tunneling site (without a donor electron), the potential energy increases when the nitrogen atom points in the direction of the field, as in Fig. 2-1. The field acts to narrow the distance between the plane of the hydrogen atoms and the nitrogen, thus increasing the frequency and energy in well q_{1B}. The opposite occurs in well q_{1A} with the energy and frequency decreasing. This frequency modulation by the electric field can occur in either mode (symmetric or antisymmetric). The NH_3 group without the donor electron can switch between modes by absorbing or releasing a photon having an energy ΔE_0.

According to the electron gating model, and to have agreement with Fig. 2-1 and have amplification (Eq. 2.31), two things must happen when the donor electron is present at an NH_3 tunneling site.

1) The electric field must be able to produce mode switching, so that the electron has an increased probability to be in the high energy vibrational mode when the nitrogen atom is pointing in the direction of the electric field as in Fig. 2-1, and an increased probability to be in the low energy vibrational mode when it flips and points away from the field. The mode switching causes a change of 180 degrees in the relative phase of oscillation in the

two energy wells. The addition of the electron to the tunneling site would essentially convert the interaction with the external electric field from frequency modulation to phase modulation. The probability for the electron being in a particular mode in one NH_3 energy well (q_{1A} or q_{1B}) depends on the magnitude and direction of the electric field.

2) The electric field must also cause a modulation of the probability for finding the donor electron orbit shifted towards the NH_3 energy well q_{1A} or towards q_{1B}. Thus, (as in Sections 2-3 and 2-4) there is a requirement to have both a change in displacement probability and an energy change for the tunneling electron, with a change in electric field across the NH_3 tunneling sites.

2-6. A voltage dependent amplification factor

Up until now the amplification has been considered constant (Eq. 2.31). Having a constant h_w is convenient for the rate and energy equations and gives results in agreement with experiment, within the normal ion channel voltage range. However, we have not taken into account the effects of the electric field interacting with the NH_3 dipole moment and the donor electron. As indicated by Eq. 2.34, there is an increase in energy above the ΔE_0 value caused by the dipole moment interacting with the electric field. This is a well-known interaction with the electric field (Feynman, 1965; Townes and Schawlow, 1955). However, for voltage dependent amplification, we need to know what happens when a donor electron is docked in the NH_3 cavity. The amplification is dependent upon the dipole moment and the displacement $2r_d$ of the donor electron orbit. Because of the complexity of the interactions, there was little hope of developing an accurate detailed theoretical model, but an overall equation for describing voltage sensitive amplification was thought to be a reasonable objective.

Since the NH_3 inversion was sensitive to an electric field, as shown by Eq. 2.34, it was assumed that a multiplying factor having the same form could represent the voltage sensitivity for amplification. To insure a uniform attenuation for various parameter values in ($h_w - 1$), this term would need to be part of the voltage dependent factor. To account for the amplification, as described in Fig. 2-1, there would also need to be a voltage-dependent mode-probability term (P_m). Incorporating these terms into a form similar to Eq. 2.34 and then combining this with Eq. 2.31 gave a reasonably simple equation:

$$h_w(v) = 1 + \frac{h_w - 1}{\sqrt{1 + \left(\frac{(h_w - 1)P_m e_0 v}{2kT \ln(h_w)}\right)^2}} \quad , \qquad (2.35)$$

and after expanding the terms

$$h_w(v) = 1 + \frac{2\beta_0 \sqrt{U} r_d^2 kT}{r\Delta E_0 \sqrt{1 + \left(\frac{\beta_0 \sqrt{U} r_d^2 P_m e_0 v}{\Delta E_0 r \ln(h_w)}\right)^2}} . \qquad (2.36)$$

When the tunneling voltage is large, all of the terms in the non-voltage dependent $(h_w - 1)$ except kT are then canceled out. The term $\ln(h_w)$ was needed to cancel the same term in the factor a_r (Eq. 2.43) for conversion to energy. The voltage-sensitive amplification factor limits the range for the rate constants, much like the finite-range rate constants developed in Chapters 4 and 6.

A missing link in the understanding of amplification was an equation for the probability of the donor electron to be in the high-energy (antisymmetric) or the low-energy (symmetric) vibrational mode. This equation would also control the probable displacement for the electron orbit resulting from an external electric field. The equation should be consistent with the description in Fig. 2-1A-C and compatible with Eq. 2.36. The equation that met this criteria, gave agreement with experimental data, and was also the most intuitive is:

$$P_m(v) = \frac{1}{1 + \exp\left(-\frac{\sqrt{2}\beta_0 \sqrt{U} r_d^2 e_0 v}{\Delta E_0 r \ln(h_w)}\right)} - \frac{1}{1 + \exp\left(\frac{\sqrt{2}\beta_0 \sqrt{U} r_d^2 e_0 v}{\Delta E_0 r \ln(h_w)}\right)} . \qquad (2.37)$$

P_m is referred to as *differential mode-probability*. Compatibility required that all of the terms present in Eq. 2.36 be also present in Eq. 2.37. If this is not the case, then changing the value of the odd parameter will distort the shape of the curve. An additional $\sqrt{2}$ term is needed to give the expected calibration. One criterion used for calibration was that the span for the amplifying range between the midpoint and saturation be equal to h_w (Fig. 4-2, Fig. 4-4). The curve for P_m has a sigmoidal shape, going from minus one to plus one for P_{m1B} in the q_{1B} energy well, and from plus one to minus one for P_{m1A} in the q_{1A} well (Fig. 2-3A). The interpretation

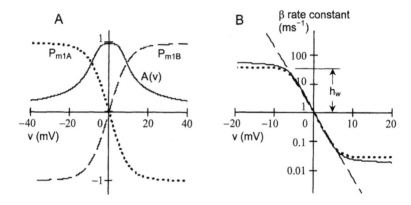

Fig. 2-3. The voltage sensitive amplification characteristic $A(v)$ determines the region over which amplified electron tunneling can occur. The differential mode-probability curve P_{m1B} corresponds to tunneling from well q_{1B}. For well q_{1A}, the curve is P_{m1A}. Either curve causes the amplification to remain constant at the maximum value over a tunneling voltage range of about -4 mV to $+4$ mV. The amplitude characteristic $A(v)$ causes the β rate constant to saturate at a factor of h_w removed from the midpoint rate (Fig. B).

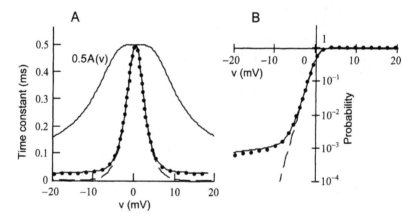

Fig. 2-4. The amplification $h_w(v)$ has an amplitude variation with tunneling voltage, shown by the curve for $0.5A(v)$, and rate-constant curves that saturate, as they approach a line with decreased slope. The time constant and probability curves for a 2-site electron-tunneling model (solid line) are in agreement with the curves (dotted line) developed in Chapters 4 and 6 for finite-range rate constants. Both have exponential equations that limit amplification to a narrow voltage range. This is in contrast with the unbounded exponential curves for a constant h_w (dashed line).

is that P_m is the probability for the electron to be in the high-energy antisymmetric mode, minus the probability to be in the low-energy symmetric mode. This would follow the description in Fig. 2-1A-C for energy well q_{1B}. The voltage-sensitive amplitude characteristic $A(v)$ for amplification is shown in Fig. 2-3A and is given by the general form

$$A(v) = \frac{1}{\sqrt{1 + \left(\dfrac{B}{1 + \exp\left(-\sqrt{2}B\right)} - \dfrac{B}{1 + \exp\left(\sqrt{2}B\right)}\right)^2}}. \qquad (2.38)$$

Voltage sensitive (or energy sensitive) amplification is given by

$$h_w(v) = 1 + \frac{\left(h_w - 1\right)a_r}{\sqrt{1 + \left(\dfrac{B}{1 + \exp\left(-\sqrt{2}B\right)} - \dfrac{B}{1 + \exp\left(\sqrt{2}B\right)}\right)^2}}. \qquad (2.39)$$

where B and a_r can take several forms:

a) tunneling voltage (v)

$$B = \frac{\left(h_w - 1\right)e_0 v}{2kT \ln\left(h_w\right)}, \quad a_r = 1 \qquad (2.40)$$

b) membrane voltage (V_m)

$$B = \frac{\left(h_w - 1\right)e_0 \eta \left(V_m - V_0\right)}{2kT \ln\left(h_w\right)}, \quad a_r = 1 \qquad (2.41)$$

c) energy (ΔE)

$$B = \frac{2\Delta E}{\Delta E_1} \qquad (2.42)$$

and

$$a_r = \frac{4kT \ln\left(h_w\right)}{\left(h_w - 1\right)\Delta E_1} \qquad (2.43)$$

where

$$h_w = 1 + \frac{2\beta_0 \sqrt{U} r_d^2 kT}{r \Delta E_0} \qquad (2.44)$$

The α_e rate constant can now be written in the following form:

$$\alpha_e(v) = \frac{K}{h_w} \exp\left[\frac{\left[(h_w-1)A(v)a_r+1\right]e_0 v}{2kT}\right]. \tag{2.45}$$

The curve for this equation is shown in Fig. 2-5 (solid line). Removing the +1 term, representing non-amplified electron tunneling, gives a curve (dotted line) that saturates at a rate-constant h_w larger or smaller that the rate constant with zero tunneling voltage.

2-7. The amplification energy window

In the electron-gating model there are two sources of energy acting to transfer charge across the tunneling sites. There are also two mechanisms for saturation.

1. The potential energy from the electric field causes the usual non-amplified charge transfer. There is no sharp saturating knee in the rate curve, but the energy curve does saturate as a result of the reduced electron probability at the donor sites. Saturation for displacement energy is described in Section 3-5.

2. The mode-switching probability change with the electric field causes amplification and saturation of the rate curve at an energy of ΔE_1. The amplification and saturation are illustrated in Fig. 2-5 and Fig. 2-7. The saturation of the electron-transfer rate was also analyzed using a second method, described in Chapter 4. This method gave a similar characteristic curve.

Up until now, the independent variable has been the tunneling voltage v or the membrane voltage V_m. To better understand the amplification, the range of free energy ΔE for the donor electron was examined. The symbol ΔE has been previously used, to represent electron free energy, in place of $-\Delta G°$ to emphasize a dependence on electrical fields (DeVault, 1984). The potential energy of the donor electron with potential v between the sites is $e_0 v$. This non-amplified tunneling term was replaced by ΔE. For the amplifying region there is a saturation potential Δv_s, which corresponds to the NH_3 energy window ΔE_1 (Fig. 2-5).

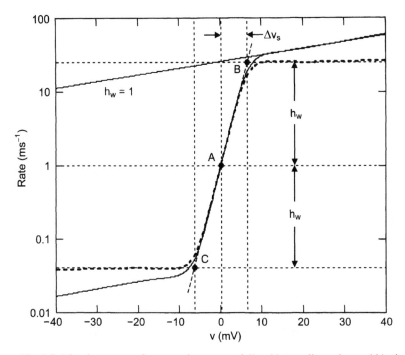

Fig. 2-5. The electron transfer rate varies exponentially with tunneling voltage within the amplifying energy window. The energy window for the NH_3 amplification is ΔE_1, which corresponds to a tunneling voltage range of $2\Delta v_s$. Outside of the saturation potential v_s, amplification decreases to unity. For the 6 Å tunneling distance with $h_w = 25.2$, the saturation voltage (point B) is $\Delta v_s = 6.4$ mV. This voltage increases with a reduction in the amplification. The curves were generated with Eq. 2.45, for rate-constant α_e, plus Eq. 2.38 and Eq. 2.40.

A relation between tunneling voltage v and the energy window ΔE_1 was developed as follows:

In Fig. 2-5, the line connecting point A and point C intersects the upper horizontal line given by h_w. The curve for h_w (solid line) includes the slope for the non-amplified rate. Here we are interested only in the amplified component, $h_w - 1$ (dotted line). The intersection of the two lines can be described as

$$\exp\left(\frac{(h_w - 1)e_0 v}{2kT}\right) = h_w \ . \tag{2.46}$$

At the point B, $v = \Delta v_s$. Then making this replacement and rearranging the terms gives

$$\Delta v_s = \frac{2kT}{(h_w-1)e_0} \ln(h_w).$$
(2.47)

An energy ratio a_r for the amplification window ΔE_1 can be defined as

$$a_r = \frac{2e_0 \Delta v_s}{\Delta E_1} = \frac{4kT}{(h_w-1)\Delta E_1} \ln(h_w)$$

$$a_r = 4.74 \quad (\text{for } h_w = 25.2, \Delta E_1 = 2.7\,\text{meV})$$
(2.48)

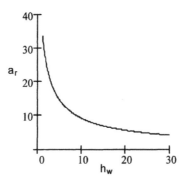

Fig. 2-6. The energy ratio for the amplification window.

Incorporating the terms for h_w (Eq. 2.44) into the equation gives

$$a_r = \frac{2r\ln(h_w)\Delta E_0}{\beta_0 \sqrt{U} r_d^2 \Delta E_1}.$$
(2.49)

Generating and interpreting the rate curves

The curves in Fig. 2-7 are for tunneling between two arginine sites with a center-to-center distance r equal to 6 Å. They were plotted from Eq. 2.45 after substituting terms from Eq. 2.38 and Eq. 2.42 through Eq. 2.44. Also, ΔE was substituted for $e_0 V$ in Eq. 2.38. These equations are based on constraints established by the model. The curves of Fig. 2-7 represent a hypothetical case, where one variable in amplification h_w (Eq. 2.44) is changed and the other factors are held constant.

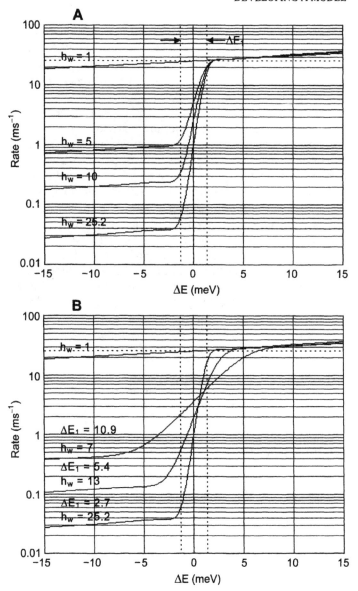

Fig. 2-7. Electron-tunneling rate curves were plotted against the effective free energy of a donor electron for several values of ΔE_1 (B). The energy window for NH_3 used in the model for amplification is $\Delta E_1 = 2.7$ meV. Outside of this window, amplification decreases to unity and the slope decreases to match that for $h_w = 1$. The ratio $\Delta E_1/\Delta E_0$ is assumed to be constant over a limited range of ΔE_0 values. If ΔE_1 is increased, then ΔE_0 must also increase and the amplification h_w must decrease (Eq. 2.44).

In a real case most, if not all, of the factors would be interdependent. The model assumes that $\Delta E_1/\Delta E_0$ is a constant. When ΔE_1 in Fig. 2-7B is increased, then ΔE_0 in h_w is also increased proportionally, thus decreasing the amplification h_w. Decreasing h_w while holding ΔE_1 and ΔE_0 constant gave the curves in Fig. 2-7A. Increasing the tunneling distance r would give similar curves, except the increase of r in the distance factor (not shown here) would shift all of the curves towards a lower rate.

One of the interesting things about the curves in Fig. 2-7 is that the rate constant is changed by a large amount with a very small change in the electron free energy. For the curve with an amplification of $h_w = 25.2$, the energy required from the electric field to increase the rate from 1 ms^{-1} to the saturation point is about 1.35 meV. This is in reasonable agreement with the maximum displacement energy calculated for transferring charge across the tunneling sites from a zero electric field condition (Eq. 3.24). When there is no amplification, Eq. 3.24 indicates that the energy from the electric field required for saturation would be about h_w times greater. Thus, the energy from the field required to transfer charge across the tunneling sites is reduced by the amplification factor. The amplification of electron tunneling reduces the input energy required for gating. This can be expressed in logarithmic form as an energy or power gain.

$$\text{Gain} = 10 \log\left(\frac{E_{d\max(1)}}{E_{d\max(25.2)}}\right) = 10 \log\left(h_w\right) = 14 \, \text{db} \qquad (2.50)$$

2-8. NH₃ inversion frequency reduction

The working hypothesis for frequency reduction is, that when the NH_3 nitrogen atom is attached to carbon and the rest of the side chain, the effective mass of nitrogen would increase, causing a reduction of its motion and resulting in a lowering of the inversion frequency. An equation for NH_3 inversion is given in Microwave Spectroscopy by Townes and Schawlow (1975) as:

$$\nu = \frac{\nu_0}{\pi} \exp\left[-\frac{2}{\hbar}\int_0^{s_0}\sqrt{2\mu(V-W)}\,ds\right], \qquad (2.51)$$

where ν is the inversion frequency, ν_0 is the vibrational frequency ($\sim 950 \text{ cm}^{-1}$), and here μ is the reduced mass of vibrational motion (not to be confused with μ for the electric dipole moment). We are interested in the inversion frequency change with a change in reduced mass; the other terms can be combined into a constant k^*.

$$v = \frac{v_0}{\pi} \exp\left(-k^*\sqrt{\mu}\right) \tag{2.52}$$

The vibrational frequency varies with the reduced mass according to the harmonic oscillator expression

$$v_0 = \frac{1}{2\pi} \sqrt{\frac{k}{\mu}} \,, \tag{2.53}$$

where k is a force constant for the vibrational frequency. Incorporating these terms into Eq. 2.52 gives

$$v = \frac{1}{2\pi^2} \sqrt{\frac{k}{\mu}} \exp\left(-k^*\sqrt{\mu}\right). \tag{2.54}$$

Taking the ratio of two inversion frequencies (f_0 and v) for two values of reduced mass (μ' and μ) gives the final equation

$$f_0 = v \sqrt{\frac{\mu}{\mu'}} \exp\left[-k_\mu\left(\sqrt{\frac{\mu'}{\mu}} - 1\right)\right]. \tag{2.55}$$

Here μ and v are the reference values for the gas phase inversion. To have agreement with the electron-gating model, the symbol f_0 is used to represent the lowered inversion frequency for an increased reduced mass μ'. An equation for the reduced mass of NH_3 is

$$\mu = \frac{3mM}{3m + M} \,. \tag{2.56}$$

This assumes that the three hydrogen atoms move together as a rigid triangle. The gas phase reference value for $N^{14}H_3$ is $v = 23.786$ GHz with $\mu = 2.487$ amu. To determine a value for the constant k_μ in Eq. 2.55, the reduced mass for $N^{15}H_3$ was calculated using the observed inversion frequency of 22.705 GHz (Townes and Schawlow, 1975). The factor k_μ was determined to be about 6.74. For a large increase in the reduced mass,

Eq. 2.55 gives only an approximate value for the inversion frequency f_0. For example, the inversion frequency for $N^{14}D_3$ was calculated as $f_0 = 2.40$ GHz versus the observed value of 1.60 GHz (Townes and Schawlow, 1975). However, when the frequency reduction is small ($f_0 > 0.5 \, v$), the calculated value for f_0 should be close to the observed value.

Calculating inversion frequency for NH_3 bonded to the carbon atom

If the mass of a carbon atom is added to the nitrogen mass (M), the reduced mass increases to $\mu' = 2.709$. This gives a calculated frequency of $f_0 = 17.0$ GHz, which is close to the Group-1 ground-state inversion frequency of 16.8 GHz, determined by microwave spectroscopy (Part II). This frequency is for a single bond between the nitrogen and carbon atoms. This would allow rotation about the bond, which could account for the observed rotational-vibrational spectra.

To account for the observed 14.3 GHz, Group-2 inversion frequency, the mass of a second nitrogen atom was added to the nitrogen and carbon for computing μ'. Adding 14.007 amu lowered the frequency f_0 to 14.6 GHz. This is in reasonable agreement with the observed value of 14.3 GHz. This suggests that the Group-2 frequency results from a double bond between the nitrogen and the carbon and that the group $H_3N=C-N$ may contribute to the reduced mass. Since rotational-vibrational spectra were present in the microwave scans for Group-2 (Fig. 10-3), some rotation about the C–N bond would seem likely.

Chapter 3

THE SETCAP MODEL

3-1. A circuit model for two-site electron tunneling

To understand the electron-tunneling response to a membrane voltage change and derive equations based on a circuit analogy, a Single Electron Tunneling Capacitor (SETCAP) model was developed. The use of this analogy was particularly helpful in developing an equation for displacement energy and an equation for the α rate constant in a model having N electron tunneling sites. In its usual form, the model represents ensemble averaged parameter values. In the circuit diagram (Fig. 3-1A), each electron-tunneling site is represented by a capacitor, which can have a maximum negative charge of 1e. When charge is displaced, it is transferred between capacitors through a tunneling resistance (R) that is inversely proportional to the tunneling probability between the sites. The tunneling resistance is constant, not varying with time or voltage. The equations for capacitance, rate constants and time constants represent steady-state values at any given voltage. The tunneling voltage v, due to the net electric field crossing the tunneling sites, is the principal independent variable in the SETCAP model.

In the circuit (Fig. 3-1A), two voltage sources ($v/2$), which vary with membrane voltage, provide the potential for maintaining a charge difference on the two tunneling-site capacitors. The exponential curves for the capacitors (Fig. 3-1B) result in a bell shaped time constant curve and sigmoidal shaped curves for the average fractional charge.

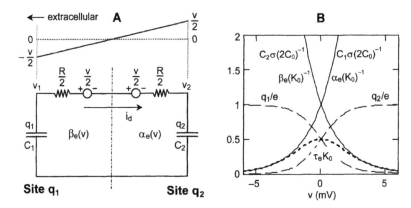

Fig. 3-1. A) SETCAP circuit for a two-site electron-tunneling model for ion channel gating. The two identical electron-tunneling sites are represented by capacitors C_1 and C_2, which together have a total negative charge of $1e$. Tunneling resistance R is inversely proportional to the electron transmission probability. Upon membrane depolarization of the sodium ion channel, a voltage is induced across the tunneling sites with the indicated polarity. This causes a displacement current i_d to flow, until a new equilibrium is reached. The displacement current has a time constant τ_e determined from the voltage-dependent rate constants $\beta_e(v)$ and $\alpha_e(v)$. **B)** The graph shows characteristic curves for fractional capacitance and fractional negative charge (q/e), for the two tunneling site capacitors. The rate coefficient $\alpha_e(K_0)^{-1}$, for site q_2, has the same curve as the fractional capacitance curve for capacitor C_1; both are unity when the potential v is zero. The time constant fraction $\tau_e K_0$ has a peak of one-half when there is equal charge on both capacitors. For the two-site model, the electron is either at the q_1 or at the q_2 site, and the ensemble averaged fractional charge is equal to the probability of finding the electron at the site. With the electron at the q_1 control site, the energy barrier for gating ion current is in the closed state. The K_0 and $(K_0)^{-1}$ multipliers cancel terms in Eqs. 3.3, 3.4 and 3.6, and create fractional values that simplify the graph scale.

These curves are different from that for the usual capacitor. Usually capacitors have a constant capacitance, with charge increasing linearly with voltage. Here, an exponential change in capacitance with voltage provides the necessary characteristic to model the charge displacement between tunneling sites with changes in membrane voltage. The capacitance characteristic for a single SETCAP capacitor is given by

$$C_n(v) = C_0 \exp\left(\frac{v_n}{H'}\right), \tag{3.1}$$

where v_n is the voltage across the capacitor, n is the site number, C_0 is the capacitance with v_n equal to zero, and H' is a sensitivity factor. With two capacitors connected in a series configuration, as in Fig. 3-1, there is a negative charge of $\frac{1}{2} e$ on each capacitor when $v = 0$. For the half of the circuit corresponding to site q_1, the time constant is given by $(R/2)C_1$ or by

$$\tau_1(v) = \frac{R}{2} \frac{2C_0}{\sigma} \exp\left(\frac{v}{2H'}\right), \tag{3.2}$$

where $2C_0/\sigma$ represents the capacitance with a negative charge of $\frac{1}{2} e$ at $v = 0$. Inverting the equation for τ_1 gives the rate constant for site q_1.

$$\beta_e(v) = \frac{\sigma}{RC_0} \exp\left(-\frac{v}{2H'}\right) \tag{3.3}$$

The rate constant at site q_2 is given by

$$\alpha_e(v) = \frac{\sigma}{RC_0} \exp\left(\frac{v}{2H'}\right). \tag{3.4}$$

A rate constant factor can be defined as

$$K_0 = \frac{\sigma}{RC_0}. \tag{3.5}$$

Adding the rate constants and inverting gives the combined time constant

$$\tau_e(v) = \frac{1}{\alpha_e(v) + \beta_e(v)} = \frac{RC_0}{\sigma} \left[\frac{1}{\exp\left(\frac{v}{2H'}\right) + \exp\left(-\frac{v}{2H'}\right)} \right]. \tag{3.6}$$

The capacitance terms of Eq 3.6 can be treated as a single time constant capacitor $C_r(v)$ representing the capacitors C_1 and C_2 in series. The term σ is a calibration factor for the time constant capacitance. It is evaluated by Eq. 3.30 based on an energy equality.

In the model, q_1 is the control site for an activation energy barrier. The probability for the barrier to be in the open state is given by

$$P_{ob}(v) = P_2(v) = \frac{\alpha_e(v)}{\alpha_e(v) + \beta_e(v)} = \frac{1}{1 + \exp\left(-\frac{v}{H'}\right)}, \tag{3.7}$$

or by the probability for the electron not to be at the q_1 control site. $P_{ob}(v) = P_2(v) = 1 - P_1(v)$. The tunneling voltage v has the same polarity as membrane voltage. Terms in the equations can be scaled to membrane voltage using the following relations: $v = \eta V_m$ and $H' = \eta H$. Since the scaling factor for v and H' is the same, we have $v/H' = V_m/H$.

3-2. Defining a capacitance factor

From the membrane voltage coefficient used in Eq. 2.16, the reciprocal sensitivity factor H is equal to

$$H = \frac{kT}{e_0 h_w \eta} \tag{3.8}$$

and H' is defined as

$$H' = \frac{kT}{e_0 h_w}. \tag{3.9}$$

Equation 3.1, for a single SETCAP capacitor, can now be written as

$$C_n = C_0 \exp\left(\frac{h_w e_0 v_n}{kT}\right), \tag{3.10}$$

and the maximum negative charge on capacitor c_n is

$$q_{max} = C_0 \int_0^{-\infty} \exp\left(\frac{h_w e_0 v_n}{kT}\right) dv$$
$$q_{max} = C_0 \frac{kT}{h_w e_0}. \tag{3.11}$$

Setting $q_{max} = e$ and solving for the capacitance factor C_0 gives

$$C_0 = \frac{e e_0 h_w}{kT}. \tag{3.12}$$

For an amplification $h_w = 25.2$, the capacitance is $C_0 = 168 \times 10^{-18}$ farads (168 aF). [Note: v_n is across each capacitor, v is between two capacitors.]

3-3. Displacement capacitance

For a small change in membrane voltage, there is a small change in voltage across the tunneling sites, which causes a displacement current and a transfer of charge. The time constant capacitance $C_\tau(v)$ determines the time rate of the charge transfer and the displacement capacitance $C_d(v)$ determines the amount of charge transfered for an incremental change in voltage. The displacement capacitance is given by

$$C_d(v) = e\frac{dP}{dv}. \tag{3.13}$$

Substituting terms for probability from Eq. 3.7 into Eq. 3.13 and taking, the derivative gives

$$C_d(v) = \frac{ee_0 h_w}{kT} \frac{1}{\left[\exp\left(\frac{h_w e_0 v}{2kT}\right) + \exp\left(-\frac{h_w e_0 v}{2kT}\right)\right]^2}. \tag{3.14}$$

The peak displacement capacitance is given by

$$C_{dp} = \frac{ee_0 h_w}{kT}\frac{1}{4} = \frac{C_0}{4}. \tag{3.15}$$

3-4. Time-constant capacitance

For a single capacitor C_τ, representing the series capacitance of the two tunneling sites and shunted by a resistance R, the capacitance can be determined by integrating the displacement current resulting from a change in voltage across the capacitor,

$$C_\tau = \frac{1}{\Delta v'}\int_0^\infty i_d dt. \tag{3.16}$$

Substituting terms for the decaying displacement current

$$C_\tau(v) = \frac{1}{\Delta v'}\int_0^\infty \frac{\Delta v'}{R}\exp\left(-\frac{t}{\tau(v)}\right)dt, \tag{3.17}$$

then solving the integral gives

$$C_\tau(v) = \frac{1}{R}\tau_e(v). \tag{3.18}$$

Substituting terms from Eq. 3.6, then replacing terms with terms from Eq. 3.9 and Eq. 3.12 gives the steady-state time-constant capacitance for the two-site model

$$C_\tau(v) = \frac{ee_0 h_w}{\sigma kT} \frac{1}{\exp\left(\frac{h_w e_0 v}{2kT}\right) + \exp\left(-\frac{h_w e_0 v}{2kT}\right)}, \quad (3.19)$$

with a peak capacitance of

$$C_{\tau p} = \frac{ee_0 h_w}{\sigma kT} \frac{1}{2} = \frac{C_0}{\sigma} \frac{1}{2}. \quad (3.20)$$

3-5. Displacement energy

When the voltage across a capacitor is altered by a small amount, the change in potential energy of the capacitor is given by the equation $dE = Cvdv$. For the two-site model the displacement capacitance is

$$C_d(v) = e\frac{dP}{dv} = \frac{C_0}{\left[\exp\left(\frac{h_w e_0 v}{2kT}\right) + \exp\left(-\frac{h_w e_0 v}{2kT}\right)\right]^2}. \quad (3.21)$$

Using this, the energy required to transfer the ensemble average charge between the two capacitors from a zero reference voltage is given by

$$E_d(v) = 2C_0\left(\frac{e_0}{e}\right) \int_0^v \frac{v}{\left[\exp\left(\frac{h_w e_0 v}{2kT}\right) + \exp\left(-\frac{h_w e_0 v}{2kT}\right)\right]^2} dv. \quad (3.22)$$

A factor of two is included to account for the change in potential energy at both the q_1 and q_2 sites. The ratio of (e_0/e) is for energy to be expressed in eV.

The integral in Eq. 3.22 can be solved using the method of integration by parts. This gives the equation

$$E_d(v) = \frac{2kT}{h_w}\left[\ln\left(\frac{2}{1+\exp\left(-\frac{h_w e_0 v}{kT}\right)}\right) - \frac{\frac{h_w}{kT}e_0 v}{1+\exp\left(\frac{h_w e_0 v}{kT}\right)}\right]. \quad (3.23)$$

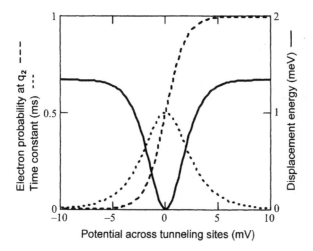

Fig. 3-2. Curves for the two-site capacitor model are shown for an amplification of $h_w = 25.2$. The displacement energy is the total energy change required to transfer charge from the $v = 0$ reference potential. It reaches a maximum because less charge is transferred as the potential (v) across the capacitors increases. The energy change required for transferring an average charge of one electron between the two capacitors is twice the maximum displacement energy.

The saturating value of $E_d(v)$, for a large positive or negative voltage, is given by Eq. 3.23, which reduces to:

$$E_{dmax} = \frac{2kT}{h_w} \ln 2 . \qquad (3.24)$$

The energy calculated with Eq. 3.23 was plotted in Fig. 3-2 using an amplification of $h_w = 25.2$. As the potential increases from $v = 0$, the energy required to transfer charge approaches a plateau where a further increase in voltage causes a negligible increase in energy. At this point almost all of the charge has been transferred. The energy change required to transfer the average charge of $1e$, from site q_1 to q_2 is twice the maximum displacement energy calculated from a $v = 0$ reference. For a voltage cycle (v_1 to v_2 back to v_1), all of the displacement energy is dissipated in the resistance R. If the amplification h_w is reduced to unity (Eq. 3.24), the displacement energy reaches a saturation that is h_w times higher. The corollary to this is that the displacement energy for electron tunneling and for voltage gating is reduced by the amplification factor.

A second method for calculating E_{dmax}

In addition to the above potential energy calculations, a second energy calculation was made for energy dissipated in the tunneling resistance. When the capacitors are discharged through the resistance R, the change in stored energy must be equal to the energy dissipated. The energy dissipated can be calculated with the well-known circuit equation

$$E_{dmax} = \int_0^\infty i^2 R\, dt .$$ (3.25)

A large hyperpolarizing voltage will cause essentially all of the charge to be in site q_1 (Fig. 3-1A). At $t = 0$, the voltage is stepped to zero, initiating a displacement of charge. The average charge at site q_1 is then given by

$$q_1 = \frac{e}{2}\left[1 + \exp\left(-\frac{t}{\tau}\right)\right].$$ (3.26)

Taking the time derivative gives the displacement current

$$i_d = \frac{dq_1}{dt} = \frac{e}{2\tau}\exp\left(-\frac{t}{\tau}\right).$$ (3.27)

Substituting these terms into Eq. 3.25 and integrating we get

$$E_{dmax} = \int_0^\infty \frac{e^2}{4\tau^2}\exp\left(-\frac{2t}{\tau}\right)R\, dt = \frac{e^2}{4}\frac{R}{2}\left(\frac{1}{\tau}\right).$$ (3.28)

Incorporating time constant parameters for $v = 0$ and a factor e_0/e to have units in eV, gives

$$E_{dmax} = \frac{e^2}{4}\frac{R}{2}\left(\frac{2}{R}\frac{\sigma kT}{ee_0 h_w}2\right)\left(\frac{e_0}{e}\right)$$

$$E_{dmax} = \frac{\sigma kT}{2h_w}$$ (3.29)

The above equation must be equal to Eq. 3.24, therefore

$$\sigma = 4\ln 2 = 2.77 .$$ (3.30)

Thus, based on the equality between the stored and dissipated energy a value for the capacitance calibration factor σ has been determined.

3-6. Energy well depth

When the tunneling voltage is changed from $v = 0$ to a saturating value, the displacement energy given by the capacitor model is E_{dmax}. Combining Eqs. 2.28 and 2.29 gives the relation

$$E_{dmax} = \frac{\Delta E_1 - \Delta E_0}{2}.$$ (3.31)

Rearranging terms for Eq. 3.24 gives

$$h_w = \frac{2kT \ln 2}{E_{dmax}}.$$ (3.32)

Substituting terms for $k_f r$ from Eq. 2.23 into Eq. 2.15 and then equating this to Eq. 3.32 gives

$$1 + \frac{2\beta_0 \sqrt{U} r_d kT}{E_{dmax}} = \frac{2kT \ln 2}{E_{dmax}}.$$ (3.33)

Substituting terms for E_{dmax} from Eq. 3.31 into Eq. 3.33, then rearranging the terms gives an equation for the energy well depth.

$$U = \left[\frac{\ln 2}{\beta_0 r_d} \left(1 - \frac{\Delta E_1 - \Delta E_0}{4kT \ln 2} \right) \right]^2$$ (3.34)

The distance r_d was already calculated as 0.27 Å with Eq. 2.24. For the values of $\Delta E_1 = 2.7 \times 10^{-3}$ eV, $\Delta E_0 = 5.93 \times 10^{-5}$ eV, Eq. 3.34 gives a well depth of about 5.75 eV.

The above equations are for a combined analysis based on the circuit model, the displacement (molecular spring) model and the energy levels of NH_3. Estimated parameter values are summarized in Table 3-1 for the two experimentally determined inversion frequencies, using equations from this section and from Chapter 2.

Table 3-1. Parameter values for the two inversion frequencies

Symbol	Group-2 inversion frequency	Group-1 inversion frequency
Group	$H_3N=C-N$	H_3N-C
f_0	14.3 GHz	16.8 GHz
f_1	650 GHz	760 GHz
ΔE_0	5.93 x 10^{-5} eV	6.95 x 10^{-5} eV
ΔE_1	2.70 meV	3.15 meV
E_{dmax}	1.32 meV	1.54 meV
h_w	25.2	21.6
x_p	42.5 Å	36.3 Å
r_d	0.27 Å	0.27 Å
η	0.106	0.124
U	5.75 eV	5.70 eV
C_0	168 aF	144 aF
C_{dp}	42 aF	36 aF
$C_{\tau p}$	30 aF	26 aF

Amplification and capacitance values are for 6°C and $r = 6$ Å.

Arginine is the amino acid used for the q_1 control sites and the adjacent q_2 sites for sodium activation, inactivation, and potassium activation gates. Based on the microwave spectra in Part II, arginine apparently has one NH_3 group inverting at the Group-1 frequency and a second NH_3 group inverting at the Group-2 frequency.

In the arginine calculations, parameter values for the Group-2 inversion frequency were used because they gave a greater amplifying range for the rate constants and a greater range for the time constant before saturation occurred. This resulted in a better match with the Hodgkin-Huxley data in the region near saturation. However, the improvement was small and electron tunneling between arginine sites might also be occurring for the Group-1 inversion. Then, parameter values for Group-1 or parameter values between Group-1 and Group-2 might be appropriate. Parameter values for the Group-1 inversion frequency would be used for electron tunneling between lysine amino acids. Tunneling between lysine and arginine would most likely use the arginine Group-1 since the energy levels are the same.

3-7. The SETCAP model for *N* tunneling sites

To determine the rate constant $\alpha_e(v)$ for a 4-site model (Fig. 3-3), a simplification is first made by combining all the tunneling site capacitors except C_1 (the control site capacitor) to form an equivalent capacitor. In steady state, there is no net current through the resistors and they can be treated as having zero resistance. Thus, in steady state, all of the capacitors are in parallel with a voltage source between them. The voltage sources between the capacitors are taken into account by the exponential terms representing the capacitors. The capacitances are added together to get the total capacitance and then divided by $N-1$ to get an equivalent capacitance C_α. The more general form $N-1$ is used, instead of 3, to allow expansion of the equation for an N site model.

$$C_\alpha = \frac{C_2 + C_3 + C_4}{N-1} \tag{3.35}$$

Substituting in terms from Eq. 3.1, gives

$$C_\alpha = \frac{C_0}{N-1}\left[\exp\left(\frac{v_2}{H'}\right) + \exp\left(\frac{v_3}{H'}\right) + \exp\left(\frac{v_4}{H'}\right)\right] \tag{3.36}$$

The rate constant α_e is equal to σ/RC_α and K_0 is defined as σ/RC_0 (Eq. 3.5). Inverting the equation for C_α, then replacing $1/C_0$ with K_0 to establish a rate constant for the equivalent capacitor C_α gives

$$\alpha_e(v) = \frac{K_0(N-1)}{\exp\left(\frac{v_2}{H'}\right) + \exp\left(\frac{v_3}{H'}\right) + \exp\left(\frac{v_4}{H'}\right)}. \tag{3.37}$$

Multiplying the numerator and denominator by $\exp(-v_1/H')$, gives the equation

$$\alpha_e(v) = \frac{K_0(N-1)\exp\left(-\frac{v_1}{H'}\right)}{\exp\left(-\frac{v_1-v_2}{H'}\right) + \exp\left(-\frac{v_1-v_3}{H'}\right) + \exp\left(-\frac{v_1-v_4}{H'}\right)}. \tag{3.38}$$

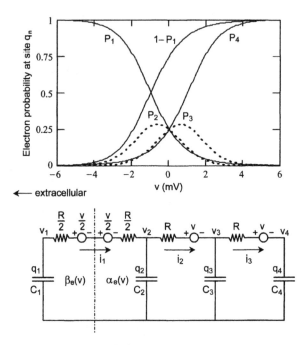

Fig. 3-3. A SETCAP circuit model for four arginine electron-tunneling sites spaced every third residue. The potential v across adjacent tunneling sites is determined by the net electric field, which varies with the membrane voltage. With $v = 0$, the average negative charge on each capacitor is $\frac{1}{4} e$. Site q_1 is defined as the control site with a rate constant $\beta_e(v)$. The capacitors for the other three sites are combined into a single equivalent capacitor and represented by rate constant $\alpha_e(v)$. The graph shows the steady-state probability of finding the electron at each tunneling site, as the potential v is varied. The $1-P_1$ curve corresponds to the open-state probability for a sodium channel activation energy barrier. It has a one-half amplitude offset of $v_{1/2} = -1.2$ mV, due to the distribution of charge for $N = 4$.

Replacing the term v_1/H' in the numerator with the balanced circuit equivalent $v/2H'$; then, replacing the capacitor voltage terms in the denominator with the potential v across the tunneling sites, gives the equation

$$\alpha_e(v) = \frac{K_0(N-1)\exp\left(-\dfrac{v}{2H'}\right)}{\exp\left(-\dfrac{v}{H'}\right) + \exp\left(-\dfrac{2v}{H'}\right) + \exp\left(-\dfrac{3v}{H'}\right)}. \qquad (3.39)$$

Multiplying numerator and denominator by $\exp(v/2H')$ gives

$$\alpha_e(v) = \frac{K_0(N-1)}{\exp\left(-\dfrac{v}{2H'}\right) + \exp\left(-\dfrac{3v}{2H'}\right) + \exp\left(-\dfrac{5v}{2H'}\right)}. \tag{3.40}$$

Rewriting this equation in a more compact form, then adding terms for calibration coefficients c_i and a charge reduction factor $1/(N-1)$, gives a final equation for the rate constant $\alpha_e(v)$. The calibration coefficients allow corrections in the shape and sensitivity of the curve, permitting a closer match to experimental data. Each of the indexed c_i terms must have an assigned value. The charge reduction factor keeps the terms near unity.

$$\alpha_e(v) = \frac{K_0}{\dfrac{1}{N-1} \displaystyle\sum_{i=2}^{N} \exp\left[-\left(\dfrac{2i-3}{N-1}\right) c_i \dfrac{v}{2H'}\right]} \tag{3.41}$$

The corresponding equation for the rate constant $v/2H'$, for the C_1 site, is given by

$$\beta_e(v) = K_0 \exp\left(-\frac{v}{2H'}\right). \tag{3.42}$$

The rate constant factor K_0 is defined by Eq. 4.10 for a temperature of 6°C.

$$K_0 = \frac{K_{max}}{h_w} \exp\left(-\frac{\lambda}{4kT}\right) \exp\left[-\beta_0 \sqrt{U}\,(r - r_0)\right]$$

$$K_0 = 1.0 \times 10^3 \text{ s}^{-1} \tag{3.43}$$

Because the value for K_0 is equal to 1.0 ms^{-1} at 6°C, the equations in Table 6-1 can be conveniently expressed, like Hodgkin-Huxley equations, as rate coefficients without the K_0 factor. A correction must be made for other temperatures, or Eq. 4.12 can be used for K_0. The above equations for $\alpha_e(v)$ and $\beta_e(v)$ are for electron tunneling with N sites and without ion channel γ-distortion. These equations were scaled to match the sodium channel rate coefficients α_m and β_m by replacing v with $(V_m - V_{1/2})$, and H' with H, where $H = 9$ and $V_{1/2} = -35$ mV. In addition, terms from Eq. 3.8 can be substituted for H and the equations expressed as in Table 6-1C.

The tunneling voltage v, in the circuit model and the above equations, does not have a $v_{1/2}$ representing the external forces, since v represents the net potential across adjacent tunneling sites due to the vector sum of all the electric fields. Fields from the sodium ion at the gating site and fields from other tunnel track electrons contribute to the one-half amplitude ($V_{1/2}$) term that subtracts from membrane voltage. To have a close match with the Hodgkin-Huxley equation for α_m, it was necessary to alter (from unity) the sensitivity factors of the exponential terms by setting the calibration coefficients c_2, c_3 and c_4 to the values indicated in Table 6-1. This correction is most likely needed because of differences in the Coulomb force on the electron at different locations.

Good agreement with the Hodgkin-Huxley equations and data for sodium activation was obtained with $N = 4$ or $N = 5$, while $N = 3$ gave poor agreement. The value of N equal four or five is also in agreement with the amino acid sequence determined by Rosenthal and Gilly (1993) for the sodium channel of the giant axon of the squid *Loligo opalescens*. It shows four arginine or lysine amino acids in domain I and five in domain II and III, spaced at every third residue of the S4.

Charge reduction factor

Ideally, the calibration coefficients (for Eq. 3.41) should be equal to one. This would require that the distribution of charge at the tunneling sites be taken into account, since the voltage sensitivity depends on the amount of charge at the site. In the term $(2i-3)/(N-1)$, the division by $N-1$ reduces the charge at each site as N increases. Without any distorting fields at ($v = 0$), the charge at each tunneling site is $(1/N)e_0$; however, the q_1 site is associated with rate constant β_e. Removing this site leaves the term $[1/(N-1)]e_0$ for the rate constant α_e. For a two-site model $N = 2$ and Eq. 3.41 is then reduced to $\alpha_e(v) = K_0 \exp(v/2H')$.

Chapter 4

AMPLIFIED ELECTRON TUNNELING AND THE INVERTED REGION

4-1. Amplification and the Marcus inverted region

Electron tunneling across proteins has been shown experimentally to approximately follow a gaussian curve when rate is plotted against the free energy. When a log scale is used for the vertical axis, the curve is an inverted parabola. The equation for the Marcus inverted region (Marcus and Sutin, 1985) is frequently used to characterize data from electron-tunneling experiments. When the free energy ΔG° equals the reorganization energy λ, the curve is at the peak rate and $K_{ET} = K_w$.

$$K_{ET} = K_w \exp\left(\frac{-(\Delta G^\circ + \lambda)^2}{4\lambda kT}\right) \qquad (4.1)$$

The factor K_w is given by

$$K_w = K_{max} \exp\left[-\beta(r - r_0)\right] \qquad (4.2)$$

For electron tunneling between the two energy wells, the edge-to-edge tunneling distance would be $r - r_0$. With NH_3 inversion, however, there

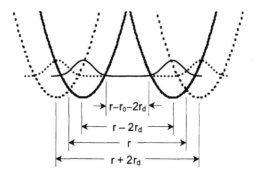

Fig. 4-1. Energy wells for adjacent donor and acceptor electron tunneling sites are geometrically displaced with the inversion of NH_3. With an inversion at both sites, the energy barrier has a smaller width ($r - r_0 - 2r_d$) for part of the inversion cycle, resulting in a greater probability for electron tunneling during this phase.

are two tunneling geometries, as shown in Fig. 4-1. The donor and acceptor energy wells are each shifted by a distance $2r_d$ with the inversion of NH_3. For a balanced condition ($\Delta G° = 0$), there is an equal probability for finding the electron in either of the two oscillating energy wells (Fig. 4-1). The intense electric field from the donor electron ($\sim 10^9$ V m^{-1}) at the adjacent acceptor dipole likely acts to minimize the electron tunneling distance for one-half the inversion cycle. The edge-to-edge distance between the electron energy wells is then ($r - r_0 - 2r_d$). For the other one-half of the inversion cycle the distance is ($r - r_0 + 2r_d$). The $4r_d$ increase in distance reduces tunneling probability by more than an order of magnitude. Thus, electron tunneling would have the highest probability for occurring during the part of the inversion cycle that has the smallest tunneling distance ($r - r_0 - 2r_d$). The increase in tunneling probability by the displacement $2r_d$ is accounted for by the r_d term in the amplification factor. The derivation of Eq. 2.14 indicated this by only having $r - r_0$ in the distance factor.

In Eq. 4.2, the $K_{max} = 10^{13}$ s^{-1} value is the experimentally determined electron-transfer rate across proteins at van der Waals contact (Page et al., 1999; Moser et al., 1992). The theory for the K_{max} rate is described elsewhere and here we will use the accepted experimental value. The maximum electron transfer rate at the 6 Å tunneling distance depends on the value used for β. For low free energy ($<2kT$) and a space jump, the value was determined from Eq. 2.4 as: $\beta = \beta_0 \sqrt{U} = 2.458$. This gives a value of $K_w = 7.33 \times 10^{10}$ s^{-1} (Eq. 4.2). This value was used for plotting electron transfer rate versus energy curves.

Electron tunneling across arginine or lysine sites on the S4 segment of ion channel proteins should follow the inverted parabola curve at high free energy. The problem was how to incorporate the steep exponential rate constant of amplified electron tunneling with the equation for the Marcus inverted region. After some mathematical manipulations, an equation was found having the desired properties.

$$K_{ET} = \frac{K_w \exp\left(\dfrac{-(\Delta G^\circ + \lambda)^2}{4\lambda kT}\right)}{1 + \left(h_w - 1\right)\dfrac{h_w}{1 + \left(h_w - 1\right)\exp\left(\dfrac{-\rho\left(h_w - 1\right)a_r \Delta G^\circ}{2kT}\right)}} \tag{4.3}$$

When the free energy ΔG° is zero, the rate constant is reduced by the amplification h_w. When h_w equals unity, the equation reduces to the Marcus expression. Equation 4.3 was used in plotting the graph for Fig. 4-2. Equation 4.4, combining forward and reverse tunneling was used in plotting the graph for Fig. 4-3.

$$K_{ET} = \frac{K_w \exp\left(\dfrac{-(\Delta G^\circ + \lambda)^2}{4\lambda kT}\right)}{1 + \dfrac{h_w\left(h_w - 1\right)}{1 + \left(h_w - 1\right)\exp\left(\dfrac{-\rho\left(h_w - 1\right)a_r \Delta G^\circ}{2kT}\right)}} + \frac{K_w \exp\left(\dfrac{-(-\Delta G^\circ + \lambda)^2}{4\lambda kT}\right)}{1 + \dfrac{h_w\left(h_w - 1\right)}{1 + \left(h_w - 1\right)\exp\left(\dfrac{\rho\left(h_w - 1\right)a_r \Delta G^\circ}{2kT}\right)}} \tag{4.4}$$

The factor $\rho = 1.14$ is used to compensate for the loss of slope-sensitivity h_w in the equations. The factor a_r calibrates the narrow amplifying region of the curve for energy. This calibration is only apparent with an expansion of the energy scale as shown in Fig. 4-4. Factor a_r is needed because in the amplifying region the effective free energy is determined by the coupling between the donor electron and the NH_3 dipole moment in the presence of the electric field. This interaction produces an amplifying energy window with a tunneling voltage width of $2\Delta v_s$.

$$a_r = \frac{2\Delta v_s}{\Delta E_1} = \frac{4kT \ln\left(h_w\right)}{\left(h_w - 1\right)\Delta E_1}. \tag{4.5}$$

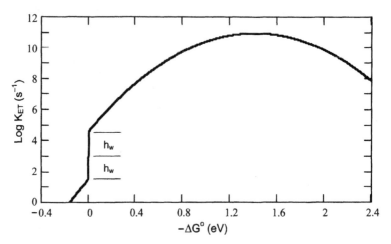

Fig. 4-2. Amplified electron tunneling from NH_3 inversion resonance at arginine donor and acceptor sites produces, according to the model, a rapid change in the electron transfer rate near zero free energy. When the amplification decreases to unity the curve merges with the curve for the Marcus inverted region. The curve is for two amplifying arginine sites on a α-helix with a 6 Å tunneling distance. It was calibrated for a rate of 10^3 s^{-1} at zero free energy (Eq. 4.3).

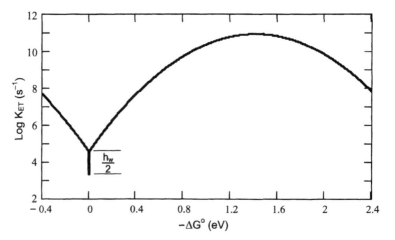

Fig. 4-3. Combining the rate for forward and reverse electron tunneling between two amplifying sites produces a narrow inverted resonant peak near zero free energy. This amplifying region of the curve corresponds to the reciprocal of the time-constant curve for the electron gating model. The curve was plotted from Eq. 4.4.

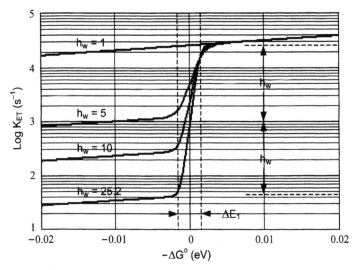

Fig. 4-4. Electron-tunneling rate curves were plotted against the effective free energy of a donor electron for several values of amplification. The energy window for NH_3 used in the model for amplification is $\Delta E_1 = 2.7$ meV. Outside of this window, amplification decreases to unity and the slope decreases to match that for $h_w = 1$. Similar curves, based on a voltage (and energy) dependent amplification factor are shown in Fig. 2-7A.

For a given value of ΔE_1, the ΔG° to reach saturation and the width of the amplifying region are constant. The slope can change because the rate at which saturation occurs varies with (h_w-1) as indicated in Figs. 4-4 and 2-7.

Some math software may need a small additional program to plot the curves in Fig. 4-2 and Fig. 4-3, so as to avoid truncation and cover the large range for K_{ET}. Before plotting the rate curve, λ was set for a rate calibration of 10^3 s^{-1} at $\Delta G = 0$ using Eq. 4.3. This gave a value of $\lambda = 1.431$ eV. Equation 4.4 was also used with $\lambda = 1.431$ eV, which gave a rate of 2000 s^{-1} at $\Delta G^\circ = 0$, corresponding to the reciprocal of the 0.5 ms time constant peak for the sodium channel. This is an unusually high value for λ and it raised a question as to how it could be about twice the typical value of 0.7 eV. Energy calculations, using the dipole moment of NH_3, indicated that removing the electron from the dipole to the midpoint between the two tunneling sites might account for up to one-half of the λ value. The energy barrier for the dipole-electron interaction would reach its maximum value midway between the two tunneling sites and this energy would be part of the activation energy barrier and increase the value for λ.

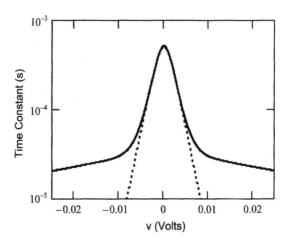

Fig. 4-5. The time constant curve for the two-site model without amplification (dotted line) decreases without limit and with a constant exponential slope. The time constant curve for amplified electron tunneling (solid line) reaches a plateau with a substantially reduced slope. This suggests that the equations of Table 6-1 have a range of validity that does not extend beyond the data of Hodgkin and Huxley. The finite-range rate constants for the electron-tunneling model (Table 6-2) match the Hodgkin-Huxley equations over the range of recorded data, but the equations diverge beyond the recorded data (Fig. 6-3).

Using Eq. 4.4 as a reference, the $\alpha(v)$ and $\beta(v)$ rate constant for the two-site electron-tunneling model can be defined to include the inverted region. The free energy ΔG° and factor a_r were replaced by $e_0 v$. Then the rate constants α and β became:

$$\alpha_e(v) = \frac{K_w \exp\left(\dfrac{-(-e_0 v + \lambda)^2}{4\lambda kT}\right)}{1 + \dfrac{h_w(h_w - 1)}{1 + (h_w - 1)\exp\left(\dfrac{(h_w - 1)\rho e_0 v}{2kT}\right)}}, \qquad (4.6)$$

$$\beta_e(v) = \frac{K_w \exp\left(\dfrac{-(e_0 v + \lambda)^2}{4\lambda kT}\right)}{1 + \dfrac{h_w(h_w - 1)}{1 + (h_w - 1)\exp\left(\dfrac{-(h_w - 1)\rho e_0 v}{2kT}\right)}}. \qquad (4.7)$$

These equations were used for computing a time constant curve (solid line) in Fig. 4-5. A conventional two-site time constant curve for the sodium activation gate is shown for comparison.

4-2. The Q_{10} temperature factor

Sodium and potassium ion channel rate constants have a basic temperature factor of about $Q_{10} = 3$ (Hodgkin and Huxley, 1952). For the electron-gating model there are at least three components that contribute to the temperature factor. The principal component is due to an energy barrier of $\lambda/4$, from the reorganization energy for electron tunneling. A smaller component can be attributed principally to ions crossing energy barriers in the channel pore. A third component may be due to a reduction in the edge-to-edge electron tunneling distance because of increased thermal vibrations.

Sodium channel pore temperature factor

According to the analysis for flux gating (Section 7.3), the temperature factor for the pore would depend on whether the channel is in the fully open state or in a region of attenuation by gating. The time required for ions to transit across the channel energy barriers (which determines the flux) depends strongly on gating. For a fully open channel, the time delay for an ion to cross the barriers is additive, and a change in transit time with temperature would be the sum of the changes in transit time for each energy barrier. When the channel is gated and the open-channel probability is small, there are narrow windows in time, when all of the gated barriers are in the open state. In these narrow windows, the probability for an ion to cross all of the barriers (with a forward driving force on the ion) is the product of the probabilities to cross each barrier. The probability of crossing an energy barrier G_0 in a narrow time window is proportional to $\exp(-G_0/kT)$ and with n barriers it would be proportional to $\exp(-nG_0/kT)$.

A factor Q'_{10} was calculated based on four gated energy barriers in the channel. It was assumed that they all have a height of G_0 equal to the calculated open-gate energy barrier of $G_{oh} = 50$ meV, for the sodium inactivation gate (Section 5-3). The equation for sodium ion current was written as

$$i_{Na}(T) = g'_{Na}\exp\left(-\frac{4G_0}{kT}\right)P_{ob}^3 \gamma_{om}(V_m - E_{Na}).$$ (4.8)

For V_m in the negative region, the γ_{om} distortion factor change with temperature is small (as a result of the discrimination in Eq. 5.26) and is neglected. The temperature change for the electron-tunneling open-barrier probability (P_{ob}) is determined separately, and P_{ob} is treated as a constant here. Taking the ratio of $i_{Na}(T)$ at two temperatures gives the temperature factor for the pore as

$$Q'_{10} = \frac{i_{Na}(T)}{i_{Na}(T')} = \exp\left[\frac{-4G_o}{k}\left(\frac{1}{T} - \frac{1}{T'}\right)\right],$$

$$Q'_{10} = 1.33 \qquad (Q'_{10})^{\frac{1}{3}} = 1.1$$

(4.9)

where T' is the reference temperature and T is the actual temperature; $T' = 273 + 6$ and $T = 273 + 16$, $G_o = 50$ meV.

Electron tunneling rate and temperature sensitivity with ΔG equal zero

A simplified equation for the electron-tunneling rate is obtained by setting the free energy terms in Eq. 4.3 equal to zero. Substituting into Eq. 4.3 the terms for K_w (from Eq. 4.2) gives K_{ET} for zero free energy, which is defined as the rate constant factor K_0.

$$K_0 = \frac{K_{max}}{h_w}\exp\left(-\frac{\lambda}{4kT}\right)\exp\left[-\beta_0\sqrt{U}\left(r - r_0\right)\right]$$

$$K_0 = 1.0 \times 10^3 \text{ s}^{-1} \text{ at } 279^\circ \text{ K}$$

(4.10)

Amplification h_w varies directly with absolute temperature, because of the kT term (Eq. 2.31). Taking the ratio of K_0 at two temperatures gives

$$Q''_{10} = \frac{K_0(T)}{K_0(T')} = \frac{T'}{T}\exp\left[\frac{-\lambda}{4k}\left(\frac{1}{T} - \frac{1}{T'}\right)\right].$$

(4.11)

For $\lambda = 1.431$ eV, $T' = 279$, $T = 289$, the calculated temperature factor is $Q''_{10} = 1.62$. Multiplying this by $[Q'_{10}]^{1/3}$ gives a temperature factor of about 1.8 for a rate constant.

Total temperature factor

In order to account for the $Q_{10} = 3$, determined experimentally, a third component was needed that did not substantially alter the rate. For this component, it was hypothesized that there would be a small reduction in the edge-to-edge tunneling distance with increasing thermal vibrations. The side chains of arginine and lysine are among the longest of the amino acids and their thermal motions are likely to contribute to the electron tunneling temperature factor. The amino acid side chains often show a considerably larger motional response to temperature than the backbone (Lee et al., 2002). With increasing temperature, the increase in thermal motions should cause the NH_3 end groups of adjacent donor and acceptor sites to map out a larger volume, thus reducing the minimum edge-to-edge tunneling distance. Based on this, an additional term was incorporated into Eq. 4.10 to represent a change in the tunneling distance with temperature.

$$K_0 = \frac{K_{max}}{h_w} \exp\left(-\frac{\lambda}{4kT}\right) \exp\left\{-\beta_0 \sqrt{U}\left[r - r_0 - r_i\left(1 - \frac{T'}{T}\right)\right]\right\} \qquad (4.12)$$

Taking the ratio of K_0 at the two previously defined temperatures, T and T', and then incorporating the temperature factor of the pore, gives an expression for the Q_{10} temperature factor for the sodium channel activation rate constant.

$$Q_{10} = \left[Q'_{10}\right]^{\frac{1}{3}} \frac{T'}{T} \exp\left[\frac{-\lambda}{4k}\left(\frac{1}{T} - \frac{1}{T'}\right)\right] \exp\left[\beta_0 \sqrt{U} r_i\left(1 - \frac{T'}{T}\right)\right] \qquad (4.13)$$

If, as a result of thermal motions, the minimum edge-to-edge tunneling distance between the adjacent donor and acceptor tunneling sites is reduced by about 0.2 Å for a 10°C temperature rise, then Eq. 4.13 would give a temperature factor of $Q_{10} = 3$. A value of $r_i = 6.15$ Å was needed to have this change with $\lambda = 1.431$ eV. There could be other components with a temperature coefficient, such as λ or K_{max}, that might contribute to Q_{10}, but these are generally small and are usually ignored.

Equation 4.13 was combined with the equation for the pore (Eq. 4.9) to give a simplified expression for determining the overall temperature factor.

$$Q_\Delta = \frac{T'}{T} \exp\left[\left(\frac{-\lambda}{4k} - \frac{4G_0}{3k}\right)\left(\frac{1}{T} - \frac{1}{T'}\right)\right] \exp\left[\beta_0 \sqrt{U} r_i\left(1 - \frac{T'}{T}\right)\right] \qquad (4.14)$$

Here Q_Δ is the factor for a change between any two temperatures T' and T.

Temperature factor for electron tunneling across intervening residues

In the electron tunneling model, the basic $Q_{10} = 3$ temperature factor, applies to channels with arginine sites spaced every third residue. When the tunneling distance increases, beyond about 10 Å the intervening non-amplifying residues become a significant factor in determining the rate constant and they most likely reduce the temperature factor. The high temperature factor for a space-jump is for the high $\lambda/4$ energy barrier and for the thermal motions of the long arginine side chains. But for non-amplifying, neutral amino acids, the temperature factor would be lower. Reducing λ to 0.7 eV, a typical value for these residues lowers the Q_{10} factor to 2.3 (Eq. 4.13). In addition, the side chains are typically shorter than for arginine. This could reduce the temperature factor for thermal motions. Reducing the factor r_i by one-half, to 3.07 Å, gives a further reduction of Q_{10} to 1.8. Thus, tunneling across the intervening residues between far sites might have a Q_{10} temperature factor in the range of 1.8 to 2.3.

4-3. Time constant

Equation 4.12 can be written for the sum of rate constants α_e and β_e for forward and reverse electron tunneling. Inverting this, gives an equation for the electron-tunneling time constant in the region of amplification.

$$\tau_e(v) = \frac{h_w}{K_{max}} \exp\left(\frac{\lambda}{4kT}\right) \exp\left\{\beta_0\sqrt{U}\left[r - r_0 - r_i\left(1 - \frac{T'}{T}\right)\right]\right\} f_\tau(v) \qquad (4.15)$$

where $f_\tau(v)$ for the two-site model is

$$f_\tau(v) = \frac{1}{\alpha_e(v) + \beta_e(v)} = \frac{1}{\exp\left(\dfrac{h_w e_0 v}{2kT}\right) + \exp\left(-\dfrac{h_w e_0 v}{2kT}\right)}. \qquad (4.16)$$

At the peak time constant, $f_\tau(v)$ is one-half and $\tau_e(v)$ is equal to 0.5×10^{-3} s for agreement with the peak time constant of sodium channel activation. Equation 4.15 can also be expressed using terms from the SETCAP model of Section 3-1. The terms h_w/K_{max} are replaced by $R_0 C_0/\sigma$.

$$\tau_e(v) = R_0 \exp\left(\frac{\lambda}{4kT}\right) \exp\left\{\beta_0\sqrt{U}\left[r - r_0 - r_i\left(1 - \frac{T'}{T}\right)\right]\right\} \frac{e e_0 h_w}{\sigma kT} f_\tau(v) \qquad (4.17)$$

4-4. Contact resistance

In the SETCAP model there is a basic resistance R_0, which is expressed in ohms. This is the resistance remaining when the edge-to-edge tunneling distance becomes zero and the energy barrier is zero. Setting Eq. 4.17 equal to Eq. 4.15 and canceling terms gives an expression for this contact resistance.

$$R_0 = \frac{\sigma kT}{ee_0 K_{max}}$$

$$R_0 = 4.16 \times 10^4 \, \text{ohms at } 279°\text{K}$$

(4.18)

4-5. Tunneling resistance

The time constant $\tau_e(v)$ can be expressed as $R(C_0/\sigma)f_\tau(v)$. Incorporating this into Eq. 4.17 leads to an equation for the tunneling resistance between adjacent space-jump tunneling sites.

$$R = R_0 \left(\frac{T}{T'}\right) \exp\left(\frac{\lambda}{4kT}\right) \exp\left\{\beta_0 \sqrt{U}\left[r - r_0 - r_i\left(1 - \frac{T'}{T}\right)\right]\right\}$$

$$R = 1.6 \times 10^{13} \, \text{ohms at } T = 279°\text{K}$$

(4.19)

This is for arginine sites with a 6 Å center-to-center tunneling distance. Temperature T' is the calibration reference temperature of 279°K.

The equations in Section 4-2 through 4-5 that have the distance factor term $\beta_0 \sqrt{U}$, are for space-jump tunneling. This is limited to the three or four residue spacing on the α-helix.

4-6. Electron tunneling site-selectivity

Why do the electrons stay in the tunnel tracks? A simple answer is that it takes less energy to transfer charge and there is a longer dwell time for the electron at the amplifying arginine and lysine sites. When h_w is decreased from 25 to 1, the energy from an external electric field required to transfer an electron to an adjacent site increases by ~25 fold (Eq. 3.24). The other consequence of amplification is that h_w increases the average dwell time, so the probability of finding an electron at a non-amplifying site is about $1/h_w$ with no electric field across the sites and with other factors being the same. In the circuit model, the increase in the dwell time is indicated by the capacitance C_0, which has h_w as a multiplying factor (Eq. 3.12).

Studies of electron tunneling across proteins have shown that much tunneling occurs at a relatively fast rate and dwell times are relatively short. Thus, the probability for finding the donor electron at these amino acids would be much smaller than at the long dwell-time arginine and lysine sites.

4-7. The amplification energy window and the inverted region

Here we show that the equations for the amplification energy window developed in Sections 2-6 and 2-7 can be combined with the equation for the Marcus inverted region to give a curve matching the electron-transfer rate curve, using the method of Section 4-1 (Fig. 4-2). From Section 2-6, the combined equation for amplified electron transfer is given by

$$K_{ET} = \frac{K_w}{h_w} \exp\left[\frac{-2\ln(h_w)A\Delta G^{\circ}}{\Delta E_1}\right] \exp\left[\frac{-\left(\Delta G^{\circ}+\lambda\right)^2}{4\lambda kT}\right] \tag{4.20}$$

where

$$A = \frac{1}{\sqrt{1+\left(\frac{-2\sqrt{2}\Delta G^{\circ}}{\Delta E_1}\right)^2 \left(\frac{1}{1+\exp\left(\frac{2\sqrt{2}\Delta G^{\circ}}{\Delta E_1}\right)} - \frac{1}{1+\exp\left(\frac{-2\sqrt{2}\Delta G^{\circ}}{\Delta E_1}\right)}\right)^2}} . \tag{4.21}$$

For comparison with the curve in Fig. 4-2, the following values were used:

$h_w = 25.2$, $K_w = 7.33$ x 10^6 s^{-1}, $\lambda = 1.431$ eV, $\Delta E_1 = 0.0027$ eV.
$(-\Delta G^{\circ} = \Delta E$ in Eq. 2.43)

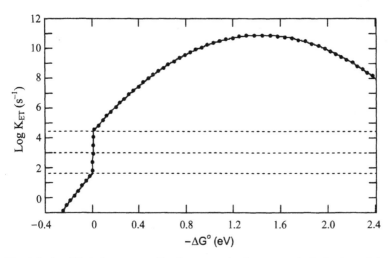

Fig. 4-6. Amplified electron tunneling is compared using two methods for determining the rate curve. The first method that was developed used Eq. 4.3 (dotted line). The second method is based on the voltage-dependent amplification factor in Section 2-6. The electron transfer rate for this method is plotted using Eq. 4.20 and Eq. 4.21 (solid line).

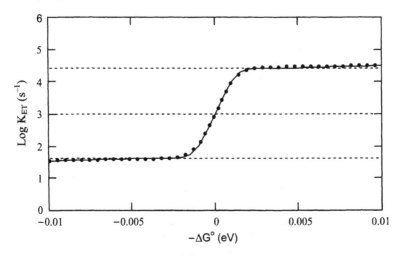

Fig. 4-7. The expanded amplifying region for both methods is in agreement. The curves are for a 6 Å center-to-center tunneling distance between amplifying arginine sites.

Chapter **5**

GATING AND DISTORTION FACTORS

In this chapter, distortions of the α and β rate constants are examined. These distortions produce the increased time constants for potassium gates and for sodium inactivation gates. Sodium activation gates have negligible distortion (below $V_{1/2}$) and the rate constants α_m and β_m are considered equivalent to α_e and β_e for electron tunneling. Most of the distortions are due to the voltage sensitivity of ion fluxes crossing energy barriers in the channel. The exception is the sodium inactivation gate, which apparently has an additional edge distortion due to the location of the tunnel track control site near the protein/cytoplasm interface.

If there were no ion channel distortions of the rate constants, α_e and β_e, then potassium gates and sodium inactivation gates would have the same peak time constant as sodium activation gates and electron tunneling. However, according to our model, the voltage sensitivity of ion transport across open-gate energy barriers in the channel can substantially modify the peak time constant and distort the shape of the time constant and open-gate probability curves. In addition, ion leakage over a single inactivation energy barrier in the closed state, distorts the open-gate probability curve and the β_h rate constant. A distortion factor γ_{ch} is derived to account for this leakage of the sodium inactivation gate. Here we consider activation and inactivation in terms of electron-modulated energy barriers; and, to be in agreement with Hodgkin and Huxley, the model for sodium activation must have three electron-modulated barriers and potassium activation must have four. Nature apparently uses multiple modulated barriers, operating in cascade for sodium and potassium gates, to substantially reduce overall ion leakage into the cell, or out of the cell for the outward rectifying potassium channels. Fortunately, sodium channel inactivation uses only a single electron-modulated barrier and the resulting distorted curve of the β_h rate constant, as described by the Hodgkin-Huxley equation, allows extraction of information about gating, such as the change in the modulated energy barrier height ΔG_h.

5-1. Sodium channel inactivation gate leakage

The Hodgkin-Huxley equations for sodium ion channel inactivation are shown in Table 6-1A. The equation for β_h does not follow a simple exponential, like the equation of β_m for activation. This deviation from a simple exponential could not be explained by electron tunneling. Instead, it was treated as ion leakage over the inactivation energy barrier, when the gate was closed. The following analysis is for the closed-gate leakage.

The total steady-state open-gate probability h is given by the probability for the gate to be in the open state, plus a term for ion leakage when the gate is closed.

$$h = P'_o + (1 - P'_o) P_L \qquad (5.1)$$

P'_o is the gate-open probability, $1 - P'_o$ is the gate-closed probability, and P_L is the voltage dependent probability of ion leakage over the barrier in the closed state. Rewriting the equation with rate coefficients α_h and β'_h, where β'_h is without ion leakage, gives

$$h = \frac{1}{1 + \dfrac{\beta'_h}{\alpha_h}} + \left(1 - \frac{1}{1 + \dfrac{\beta'_h}{\alpha_h}}\right) P_L = \frac{1 + P_L \dfrac{\beta'_h}{\alpha_h}}{1 + \dfrac{\beta'_h}{\alpha_h}}. \qquad (5.2)$$

In the voltage range with high leakage, $\beta'_h / \alpha_h \gg 1$, the equation can be written in the following form:

$$h = \frac{1}{1 + \left(\dfrac{\beta'_h}{\alpha_h}\right) \dfrac{1}{1 + P_L \left(\dfrac{\beta'_h}{\alpha_h}\right)}} = \frac{1}{1 + \left(\dfrac{\beta'_h}{\alpha_h}\right) \gamma_{ch}}. \qquad (5.3)$$

The probability of ion leakage over the barrier is given by

$$P_L = \exp\left[\frac{-\Delta G_h}{kT} + \frac{E_{Na} - V_m}{k'T}\right], \tag{5.4}$$

where ΔG_h is the change in energy barrier height between the open-gate reference state and the closed-gate state. The term, $E_{Na} - V_m$ is the driving potential across the energy barrier and the term $\exp[(E_{Na} - V_m)/k'T]$ is equal to the flux ratio for the closed-gate energy barrier. It has a value of one when zero potential is across the barrier. It is essentially the same term as developed by Ussing (1949) for independent passive transport of sodium ions. Converting Eq. 5.4 to displacement voltage by substituting $E_r - V$ for V_m, then writing an equation for distortion factor γ_{ch} based on Eq. 5.3 and Eq. 5.4, gives

$$\gamma_{ch} = \frac{1}{1 + \exp\left(-\dfrac{\Delta G_h}{kT} + \dfrac{E_{Na} - E_r + V}{k'T}\right)\left(\dfrac{\beta'_h(V)}{\alpha_h(V)}\right)^{a''}}. \tag{5.5}$$

The rate coefficient β'_h is equal to β_h without ion leakage. It has the form: $\beta'_h = 0.05\exp(-V/10)$. The other rate coefficient is $\alpha_h = 0.07\exp(V/20)$. These are Hodgkin-Huxley equations, but β'_h is without the built-in distorting factor of $\gamma_{ch} = [1 + 0.05\exp(-V/10)]^{-1}$ included in β_h. The exponent a'' is a slope correction factor that was added to allow curve matching. To be consistent with the concept of a forward coupling factor, a slightly different form for Eq. 5.5 will be used

$$\gamma_{ch} = \frac{1}{1 + \exp\left(-\dfrac{\Delta G_h}{kT} + \dfrac{E_{Na} - E_r + V}{k'T}\right)\left(\dfrac{\beta_e}{\alpha_e}\right)^{a'}\exp\left(\dfrac{-V}{4.58k'T}\right)}. \tag{5.6}$$

This is basically the same equation except that exponent a', referred to as the forward coupling factor, reduces the slope of only the gating sensitivity term β_e/α_e of electron tunneling. The multiplying exponential voltage term is a slope distortion factor, which is developed in Section 5-3. Equations for β_e and α_e are listed for Eq. 5.8. They correspond to the equations for

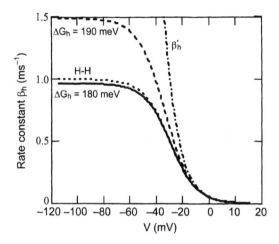

Fig. 5-1. Ion leakage over a closed inactivation gate causes distortion for the β_h rate constant. The rate constant without the ion leakage is shown as β_h'. The dotted line represents the Hodgkin-Huxley equation for β_h. The solid line represents the electron-tunneling model β_h with $\Delta G_h = 180$ meV. The curve is slightly separated for clarity. An increase in ΔG_h to 190 meV results in a substantial deviation, as shown by the dashed line. The electron-tunneling model curves were plotted from Eq. 5.7.

β_h and α_h, in Table 6-1, after removing the γ-distortion factors.

An explanation for the coupling factor a' is the following: The β_e/α_e term for electron tunneling controls the frequency of gate closures and dwell times, in response to voltage. The open-gate probability has a slope sensitivity determined by β_e/α_e, based on a completely closed gate, but the gate does not completely close the channel, and the slope sensitivity of β_e/α_e needs to be reduced because of the decrease in the span. If the force between the control site electron and the ion at the gating site were an order of magnitude greater, then the channel would be (for practical purposes) completely closed and the coupling factor a' would be unity. Leakage over the closed inactivation gate causes the slope sensitivity of h at $V_{1/2}$ to be reduced by about 6 percent, where $h = [1+\beta_h/\alpha_h]^{-1}$. The reduction is determined by computing the ratio of the slope for $h(-3$ mV) using the gate-leakage β_h/α_h and the no-leakage β_h'/α_h term. By setting $a' = 0.942$, the slope sensitivity of h at $V_{1/2}$ with no-leakage is reduced to match the slope sensitivity with gate leakage.

The equation for the rate coefficient β_h can be written as the product of the no-leakage rate constant β'_h and the gate leakage distortion factor γ_{ch}.

$$\beta_h = \frac{0.05\exp\left(\dfrac{-V}{10}\right)}{1+\exp\left(-\dfrac{\Delta G_h}{kT}+\dfrac{E_{Na}-E_r+V}{k'T}\right)\left(\dfrac{\beta_e}{\alpha_e}\right)^{a'}\exp\left(\dfrac{-V}{4.58k'T}\right)} \qquad (5.7)$$

The no-leakage rate constant β'_h can be written as the product of β_e and a slope distortion factor γ'_{oh} (Eq. 5.18). Incorporating this into Eq. 5.7 and converting to membrane voltage gives

$$\beta_h = \beta_e \gamma_{\beta h} = \frac{\beta_e b_h \exp\left(\dfrac{V_m+70}{4.58k'T}\right)}{1+\exp\left(-\dfrac{\Delta G_h}{kT}+\dfrac{E_{Na}-V_m}{k'T}\right)\left(\dfrac{\beta_e}{\alpha_e}\right)^{a'}\exp\left(\dfrac{V_m+70}{4.58k'T}\right)}, \qquad (5.8)$$

where $\alpha_e = \exp[-s(V_m - V_{1/2})/2H]$ and $\beta_e = \exp[(V_m - V_{1/2})/2H]$
$s = 0.55$ $2H = 11$

α_e and β_e rate constants were matched (same $V_{1/2} = -57$ mV) by setting the multiplying constants to $b_h = 0.06$, and shifting the unbalance of 1 mV from the β_e exponential to the distortion factor. This accounts for 10 mV of the 70 mV offset in Eq. 5.8 and β_h in Table 6-1C.

As shown in Fig. 5-1, β_h equals one when V is more negative than -100 mV ($V_m > 40$ mV). For a constant β_h, the voltage terms in the numerator and denominator of Eq. 5.8 must cancel. This occurs when a' is equal to 0.942. Setting $\beta_h = 1$ and $a' = 0.942$ in Eq. 5.8 and solving for the change in the energy barrier gives

$$\Delta G_h = -kT \ln b_h + e_0(E_{Na} - V_{1/2}) = 180 \text{ meV} . \qquad (5.9)$$

The curve for β_h (Fig. 5-1) is very sensitive to ΔG_h. Variation in this parameter (or differences in Na$^+$ concentration) could account for the almost two-to-one spread of the data points, in Fig. 9 of the Hodgkin and Huxley (1952) paper.

In addition to the closed-gate ion leakage, there is distortion of the electron-tunneling rate constants, due to the voltage sensitive sodium ion fluxes crossing the open-gate energy barriers. The following additional questions remain unanswered for the sodium inactivation gate:

1. The multiplier in the equation for β_h is only 0.05 and for α_h it is 0.07. Why are the multipliers so small?

2. The peak time constant for the sodium inactivation gate is 8.5 ms compared to 0.5 ms for the activation gate. What causes the difference?

3. The inactivation gate has a β_h sensitivity almost twice that of the sodium activation gate. Why?

Before considering these questions, the ion channel gating process for the electron-tunneling model is examined.

5-2. Ion channel gating

The following model describes ion channel gating using an electron at a tunnel-track control site as the gating agent. This model applies to both the activation and inactivation gates. It is assumed that the gating sites are nonreactive and have energy barriers with an open state, and a closed state that substantially inhibits ion flow. In the open state, the ions must cross a small open-gate energy barrier for each gating site cavity to establish current flow. The change in the energy barrier height with electron modulation is determined by the energy required to move the ion past a lateral displacement, at the gating site, against an electric field from a control site electron. It is assumed that the lateral displacement is caused by a cavity between the α-helix side chain residues in a narrow region of the channel.

When an electron tunnels to the q_1 control site, there is a change in charge Δq, which causes a change in force ΔF on the positively charged ion at the adjacent gating site cavity. The energy ΔG required to move the ion against the electric field of the control site electron, through a displacement Δr, is given by the integration of the Coulomb force over the displacement distance, r_1 to $r_1 + \Delta r$ (Fig. 5-2).

$$\Delta G = \frac{\Delta q\, e_0 10^{13}}{4\pi\varepsilon_0\varepsilon_r} \int_{r_1}^{r_1+\Delta r} \frac{1}{r^2}\, dr \qquad (5.10)$$

Thus, ΔG represents the change in energy barrier height that is modulated by the control site electron. Since the ion's advance of Δx is much smaller than the distance $r_1 + \Delta r$, the energy associated with Δx is relatively small and is neglected in the calculations.

The open-gate energy barrier G_o is due to a net force F_o, resulting from nearby aspartic (D) or glutamic (E) residues as shown in the protein sequence. In solution, these residues carry an excess negative charge, but here they act as polar residues carrying partial charges. If they carried an excess negative charge, the gate would be permanently closed. The residues are represented as r_N (Fig. 5-2, Fig. 8-1, Fig. 8-2). They can be represented approximately as dipoles with their negatively charged region oriented toward the gating site cavity. They exert a net attractive force on the ion in the cavity. This differential force is rapidly attenuated with increasing distance, so that to account for a 50 meV open-gate energy barrier, the dipoles must be within a few angstroms of an ion in the gating cavity.

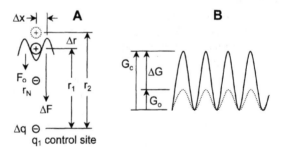

Fig. 5-2. When a cation enters a gating cavity, a net force F_o from a nearby polar amino acid residue creates an energy barrier G_o. The net force on the cation increases to $F_o + \Delta F$ when an electron tunnels to the q_1 control site. This increases the energy required for the cation to escape from the cavity, to $G_o + \Delta G$, thus closing the gate. The change in the energy barrier ΔG is determined by the displacement distance Δr, the force change ΔF and distance r_1. Illustration B shows four modulated energy barriers in the gate-closed state, each having a single open state. An energy barrier is active only when an ion is present in the gating site cavity.

Here we are interested in estimating the displacement distance Δr using the value for ΔG obtained from the sodium channel inactivation gate distortion factor. It was determined for the distortion factor γ_{ch}, that a change in the energy barrier of $\Delta G_h = 180$ meV was required to match the distortion characteristic of the Hodgkin-Huxley equation for β_h. From this energy, a displacement distance Δr was calculated by solving Eq. 5.10 and applying it in the following rearranged form:

$$\Delta r = 4\pi\varepsilon_0\varepsilon_r r_1 \left(r_1 + \Delta r\right)\frac{\Delta G}{\Delta q e_0} 10^{-13}.$$ (5.11)

The parameter values are: $\Delta G = 180$ meV, $r_1 = 8$ Å, $\Delta q = 1.6 \times 10^{-19}$ C, $\varepsilon_r = 2$, $\varepsilon_0 = 8.85 \times 10^{-12}$ $C^2N^{-1}m^{-2}$, $e_0 = 1$.

Using these values, the lateral displacement of the ion was calculated to be $\Delta r = 2$ Å. This displacement is dependent on the assumed parameter values, but it shows that a reasonable value for Δr can be obtained for a change in charge of $1e$ at the q_1 control site with a distance of $r_1 = 8$ Å between the control site and the ion. The value for Δr might range from 2 to 4 Å, depending on the assumed value for ε_r. A cavity with this depth would thus provide an open-gate to closed-gate energy barrier change of 180 meV ($7.5kT$). An open-gate energy barrier of 50 meV ($2kT$) is determined in Section 5-3. With these values, the total peak-to-valley energy for a gate is $9.5kT$. This is in agreement with the peak-to-valley energy barrier values for the sodium channel selectivity filter determined by (Hille, 1975).

The probability for an inactivation gate energy barrier to be in the open state is equal to $[1+\beta_e/\alpha_e]^{-1}$. However, the open-channel probability for the inactivation gate includes the closed-gate leakage distortion, so we get

$$P_o = \left(\frac{1}{1+\dfrac{\beta_e\gamma_{ch}}{\alpha_e}}\right).$$ (5.12)

This probability is used for flux gating in Section 7-2. The sodium channel inactivation gate also has open-gate distortion and this is part of the linear model probability given by h.

$$h = P_o\gamma_{oh} = \left(\frac{1}{1+\dfrac{\beta_e\gamma_{ch}}{\alpha_e}}\right)\gamma_{oh}$$ (5.13)

5-3. Inactivation gating and open-gate distortion

It was shown in Section 5-1 that the sodium channel inactivation gate is sensitive to the potential across the closed-gate energy barrier, because of leakage. When the energy barrier is in the open-state, there is a similar sensitivity to the ionic driving potential. The terms modulating the energy barrier are for the Ussing (1949) equation. With a strong driving potential the energy barrier, in the open state, can become saturated. To represent both modulation and discrimination resulting from saturation, the open-gate distortion is given by

$$\gamma_{oh} = \frac{N_F}{1 + \exp\left(\dfrac{G_{oh}}{kT}\right)\exp\left(\dfrac{V_m - E_{Na}}{k'T}\right)} . \tag{5.14}$$

While electron tunneling controls the frequency and duration of gate openings, the amplitude of the open-gate pulses (and the probability) is modulated by the driving potential across the channel in accordance with the modulator-discriminator function of Eq. 5.14. The term N_F is a normalizing factor that sets the amplitude for agreement with the value for h at the normalizing voltage. It is generally close to one. The term G_{oh} is for the open-gate energy barrier. Both the open-gate energy barrier and the slope sensitivity for β_e were determined by matching the membrane voltage curve for h', determined by Eq. 5.15, with a reference curve for h' from the Hodgkin-Huxley rate coefficients. The closed gate leakage γ_{ch}, which distorts the β rate constant, was removed for this match-up.

$$h' = \left(\frac{1}{1 + \dfrac{\beta_e}{\alpha_e}}\right)\gamma_{oh} = \left(\frac{1}{1 + \dfrac{\beta_e}{\alpha_e}}\right)\frac{N_F}{1 + \exp\left(\dfrac{G_{oh}}{kT} + \dfrac{V_m - E_{Na}}{k'T}\right)} \tag{5.15}$$

$E_{Na} = 55$ mV, $kT = 24$ meV

$$\alpha_e = \exp\left(-\frac{V_m + 57}{20}\right) \quad \beta_e = \exp\left(\frac{V_m + 58}{2H}\right) \tag{5.16}$$

Reference equations:

$$\alpha_h = 0.07\exp\left(-\frac{V_m + 60}{20}\right) \quad \beta_h' = 0.05\exp\left(\frac{V_m + 60}{10}\right) \quad h' = \left[1 + \exp\left(\frac{\beta_h'}{\alpha_h}\right)\right]^{-1} \tag{5.17}$$

Values determined in match-up: $N_F = 1.02$ $G_{oh} = 50$ meV $2H = 11$

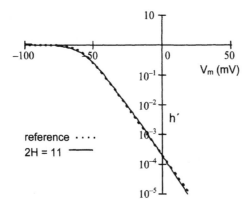

Fig. 5-3. The electron tunneling slope sensitivity for rate constant β_e and the open-gate energy barrier ΔG_h were determined by adjusting the parameters for a best fit to a reference curve. The dotted reference curve for h' is for the Hodgkin-Huxley rate constants, with distortion γ_{ch} removed. The solid line is for the electron-tunneling rate constants and the open-gate distortion factor γ_{oh}. A value of $\Delta G_h = 50$ meV and a slope sensitivity for β_e of $2H = 11$ gave agreement with the reference.

An additional match-up:

Instead of multiplying electron-tunneling probability by γ_{oh} as in Eq. 5.15, a matching curve for h' can be produced by multiplying the β_e rate constant by the distortion factor

$$\gamma'_{oh} = \exp\left(\frac{V_m + 60}{110}\right), \tag{5.18}$$

and h' then becomes

$$h' = \frac{1}{1 + \dfrac{\beta_e \gamma'_{oh}}{\alpha_e}} . \tag{5.19}$$

Equation 5.19 produces a curve, matching the reference curve in Fig. 5-3.

Determining the distortion multiplier b_h

The multipliers shown in Eqs. 5.17 for α_h and β'_h are unmatched. Matching them, by adding constants to the exponentials gives

$$\alpha_h = b_h \exp\left(-\frac{V_m + 57}{20}\right) \qquad \beta'_h = b_h \exp\left(\frac{V_m + 58}{10}\right), \tag{5.20}$$

where $b_h = 0.060$.

In Fig. 5-4 the sodium channel open-gate distortion γ_{oh} is plotted along with the electron-tunneling rate constant β_e with leakage-distortion γ_{ch}. Multiplying the curve $\beta_e \gamma_{ch}$ by the sensitivity distorting term $\exp(V_m/110)$ increases the slope, producing the curve β_{ah}.

$$\beta_{ah} = \beta_e \gamma_{ch} \exp\left(\frac{V_m}{110}\right) \qquad (5.21)$$

The electron-tunneling rate constants can be represented as a rate-of-change of flux with voltage. When $V_m = E_{Na}$, there is no ion channel current and the flux ratio is unity. At this potential, the open-gate energy barrier attenuates both fluxes equally by a factor b_o. Multiplying the curve for β_{ah} by b_o, gives the rate constant β_h.

$$\beta_h = \beta_e \gamma_{ch} \exp\left(\frac{V_m}{110}\right) b_o . \qquad (5.22)$$

For the multiplier b_h

$$\beta_h = \beta_e \gamma_{ch} \exp\left(\frac{V_m - V_{dh}}{110}\right) b_h . \qquad (5.23)$$

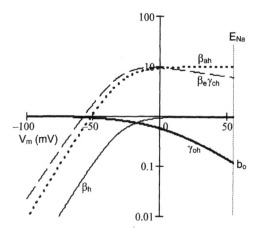

Fig. 5-4. The leakage distortion factor γ_{ch} multiplies the electron-tunneling rate constant β_e producing the curve $\beta_e \gamma_{ch}$. The open-gate distortion factor γ_{oh} increases the slope sensitivity of the open channel probability P_o (Eq. 5.13). This distortion is incorporated into rate constant β_h. Multiplying $\beta_e \gamma_{ch}$ by the voltage sensitivity term (Eq. 5.21), from the match-up, increases the sensitivity from 1/11 to 1/10 as indicated by β_{ah}. Evaluating γ_{oh} at $V_m = E_{Na}$ gives the attenuation factor b_o for the open-gate energy barrier G_{oh}. Factor b_o attenuates the curve β_{ah}, producing the final rate constant curve for β_h.

Taking the ratio of Eq. 5.21 to Eq. 5.22 gives

$$b_h = b_0 \exp\left(\frac{V_{dh}}{110}\right). \tag{5.24}$$

The factor b_0 is equal to γ_{oh} evaluated at $V_m = E_{Na}$. Terms from Eq. 5.14 are substituting for b_0. This gives the equation for b_h in terms of the open-gate energy barrier.

$$b_h = \left[\frac{N_F}{1 + \exp\left(\dfrac{G_{oh}}{kT}\right)}\right] \exp\left(\frac{V_{dh}}{110}\right) \tag{5.25}$$

In the match-up of Fig. 5-3, the value for V_{dh} is -60 mV, the open-gate energy barrier G_{oh} is 50 meV, and N_F is 1.02. For these values, $b_h = 0.065$. This is close to the 0.060 must-have value determined in Eq. 5.20. If V_{dh} is increased to -70 mV to have equal $V_{1/2}$ voltages (as in Eq. 5.8), then the calculated value from Eq. 5.25 becomes $b_h = 0.060$.

Because of the factor b_h, from open-gate distortion, the peak time constant for sodium inactivation is about 17 times longer than for sodium activation. Most of the increase in the time constants comes from the open gate energy barrier G_{oh}.

Note: The distortion factor γ'_{oh} (Eq. 5.18) is also dependent on E_{Na}, but since the distortion is small, the E_{Na} terms are included in the overall offset.

5-4. Sodium channel activation gates and distortion

The distortion curve γ_{oh} (Fig. 5-4) is the same for both sodium activation and inactivation (i.e. $\gamma_{om} = \gamma_{oh}$). The big difference for activation is that the location of the control site is opposite to where the distortion is occurring. The maximum distortion occurs when V_m equals E_{Na} and most of the charge is at the q_8 control site for sodium inactivation. For sodium activation gating, when most of the charge is at the q_1 control site, there is negligible distortion of β_e, or α_e rate constants, or the probability m' because γ_{om} remains at unity up to the $V_{1/2}$ potential. The probability m' is determined from electron tunneling probability $(1 - P_1)$ multiplied by the open-gate distortion factor (Eq. 5.26). Figure 5-5 shows agreement between m' and the reference curve m for the Hodgkin-Huxley equations, up to $V_m = -20$ mV. Since γ_{om} remains constant (at unity) below $V_{1/2}$, the

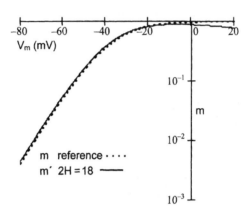

Fig. 5-5. A comparison is made between the reference Hodgkin Huxley probability m (dotted line) and the open-state probability m', for a modulated energy barrier with a distortion factor $[\gamma_{ch}]^{1/3}$ (solid line); m' is calculated by Eq. 5.26. The curve for m' shows agreement with the reference up to a membrane potential of -20 mV. The electron-tunnel rate constants β_e and α_e are considered equivalent to rate constants β_m and α_m for sodium activation.

sensitivity for the rate constant β_e must be the same as for β_m and the multiplying constant b_m should be equal (or near) one.

$$m' = \frac{1}{1+\dfrac{\beta_e}{\alpha_e}}\left(\gamma_{om}\right)^{\frac{1}{3}} = \frac{1}{1+\dfrac{\beta_e}{\alpha_e}}\left[\frac{1}{1+\exp\left(\dfrac{G_{om}}{kT}+\dfrac{V_m-E_{Na}}{k'T}\right)}\right]^{\frac{1}{3}}N_F \qquad (5.26)$$

$E_{Na} = 55$ mV, $kT = 24$ meV, $N_F = 1.1$, $G_{om} = 50$ meV, α_e and β_e have the same equation as α_m and β_m.

In Eq. 5.26, the term $\exp[G_{om}/kT+(V_m-E_{Na})k'T]$ is less than 0.36 for V_m more negative than -20 mV. This effectively discriminates against the distortion. The 50 meV for G_{om} is from the inactivation gate; the energy barrier height could not be determined here.

Based on the above, the electron-tunnel rate constants β_e and α_e are considered equivalent to rate constants β_m and α_m for sodium activation gates.

5-5. Potassium channel gating and distortion

The electron-tunneling gating model has allowed insight into previously unexplained aspects of the Hodgkin-Huxley equations for the sodium channel. The following questions need to be answered for the potassium channel:

1. The potassium ion channel equation for β_n has a very low sensitivity ($V/80$), compared to the sodium channel sensitivity ($V/18$). Why is there a large difference in sensitivity?

2. There is a 24 mV offset in the alignment of V_P for the time constant peak and the $V_{1/2}$ voltage, and both curves are skewed by more than can be attributed to multiple tunneling sites. What causes the large alignment offset and large skewing of the curves?

3. The multiplier for β_n is only 0.125. Why is the multiplier so small?

4. The peak time constant for the potassium ion channel gate is 11 times greater than for the sodium activation gate. Why?

As with the sodium inactivation gate, answers came from examining the open-gate energy barrier distortions. It is assumed that the potassium ion channel follows the Hodgkin-Huxley n^4 model. This is represented by four modulated energy barriers, each barrier controlled from a separate electron tunnel track. The Ussing flux ratio term $\exp[(V_m-E_K)/k'T]$ is used to represent the influence of membrane potential across the open-gate energy barriers. The modulator-discriminator function (Eq. 5.14) for the sodium channel was modified to have the potassium flux ratio parameters.

$$\gamma_{on} = \frac{N_F}{1+\exp\left(\dfrac{G_{on}}{kT} - \dfrac{V_m-E_K}{k'T}\right)} \tag{5.27}$$

Both the open-gate energy barrier G_{on} and the slope sensitivity for β_e were determined by matching the membrane voltage curve for n_a^4, determined by Eq. 5.28, with a reference curve for n^4 from the Hodgkin-Huxley rate coefficients. The equation for n_a^4 is

$$n_a^4 = \left(\frac{1}{1+\dfrac{\beta_e}{\alpha_e}}\right)^4 \left[\frac{N_F}{1+\exp\left(\dfrac{G_{on}}{kT} - \dfrac{V_m-E_K}{k'T}\right)}\right], \tag{5.28}$$

where $E_K = -75$ mV, $k'T = 24$ mV and

$$\alpha_e = \cfrac{1}{\cfrac{1}{N-1}\sum_{i=2}^{N}\exp\left(-\cfrac{2i-3}{N-1}c_i\,\cfrac{V_m-V'}{2H}\right)}$$

$$\beta_e = \exp\left(-\cfrac{V_m-V'}{2H}\right).$$

$$N = 7$$
$$c_2 = 1.53 \quad c_5 = 1.2$$
$$c_3 = 2.4 \quad c_6 = 1.2$$
$$c_4 = 1.7 \quad c_7 = 1.5$$

(5.29)

Reference equations:

$$\alpha_n = \cfrac{-0.01(V_m+50)}{\exp\left(-\cfrac{V_m+50}{10}\right)-1} \qquad \beta_n = 0.125\exp\left(-\cfrac{V_m+60}{80}\right)$$

$$n^4 = \left[1+\exp\left(\cfrac{\beta_n}{\alpha_n}\right)\right]^{-4}$$

(5.30)

Values determined in match-up:
$N_F = 0.94 \quad G_{on} = 54$ meV $\quad H = 13 \quad V' = -60$ mV

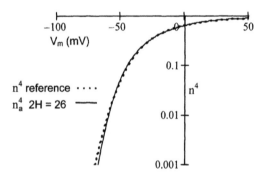

Fig. 5-6. The electron tunneling slope sensitivity for rate constant β_e, the open-gate energy barrier G_{on}, and offset V' were determined by adjusting parameters for a best fit to a reference curve. The dotted reference curve for n^4 is with the Hodgkin-Huxley rate constants. The solid line for n_a^4 is with the electron-tunneling rate constants combined with open-gate distortion factor γ_{on} (Eq. 5.28). An energy barrier $G_{on} = 54$ meV, a slope sensitivity for β_e of $2H = 26$, and an offset of $V' = -60$ mV gave reasonable agreement with the reference.

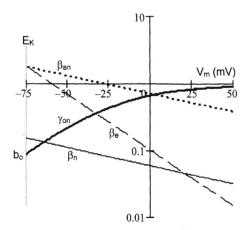

Fig. 5-7. The open-gate distortion factor γ_{on} multiplies the electron-tunneling probability (Eq. 5.28), decreasing its voltage sensitivity. This distortion can be incorporated into the β rate constant to match β_n for the linear model. Multiplying the rate constant β_e by the sensitivity distortion term (Eq. 5.31) decreases the slope from 1/26 to 1/80 as shown for β_{an}. The open-gate distortion factor also causes a voltage-independent decrease of b_o for both α and β rate constants. A curve matching β_n is obtained by multiplying the curve β_{an} with factor b_o, which is the distortion γ_{on} evaluated at $V_m = E_K$.

The equation for the multiplying factor b_n is similar to that for sodium inactivation. When $V_m = E_K$, there is no channel current and the flux ratio is unity. At this potential, the open-gate energy barrier attenuates both fluxes and rate constants equally by a factor b_o. Multiplying the curve for β_e by $\exp[(V_m - E_K)/38.4]$ rotates the curve about the $V_m = E_K$

$$\beta_{an} = \beta_e \exp\left(\frac{V_m - E_K}{38.4}\right) \tag{5.31}$$

Then, multiplying by b_o, gives the rate constant β_n.

$$\beta_n = \beta_e \exp\left(\frac{V_m - E_K}{38.4}\right) b_o \tag{5.32}$$

The multiplier b_o is scaled to b_n by including an offset term V_{dn}

$$\beta_n = \beta_e \exp\left(\frac{V_m - E_K - V_{dn}}{38.4}\right) b_n. \tag{5.33}$$

Taking the ratio of Eq. 5.32 and Eq. 5.33, gives

$$b_n = b_o \exp\left(\frac{V_{dn}}{38.4}\right). \tag{5.34}$$

Evaluating γ_{on} at $V_m = E_K$ (Eq. 5.27) gives b_o, then substituting terms into Eq. 5.34 for b_o gives

$$b_n = \left[\frac{N_F}{1 + \exp\left(\frac{G_{on}}{kT}\right)}\right] \exp\left(\frac{V_{dn}}{38.4}\right). \tag{5.35}$$

A value of $V' = -60$ mV was determined in the match-up of Fig. 5-6. The V' term represents the one-half amplitude voltage for electron tunneling, which for potassium is about 12 mV more negative than the overall $V_{1/2}$ that is shifted by distortion γ_{on}. With $V' = -60$ mV, the α_e rate constant (Eq. 5.29) must have a multiplier of $b_n = 0.059$ for agreement with the reference equation for α_n (Eq. 5.30). Substituting this value for b_n into Eq. 5.33 and substituting the exponential terms for β_e (Eq. 5.29) and β_n (Eq. 5.30), then solving for V_{dn} gives a value of -14 mV.

With an open-gate energy barrier of $G_{on} = 54$ meV and $N_F = 0.94$ from the match-up, and with $V_{dn} = -14$ mV, the calculated value from Eq. 5.33 is $b_n = 0.062$. This value is greater that the must-have value of 0.059 and it suggests that the energy barrier is higher. With $G_{on} = 55.4$ meV, the calculated value for b_n is 0.059.

The γ_n-distortion factors, including the b_n multiplier, cause most of the 24 mV alignment offset between V_p and $V_{1/2}$ and account for an 11-fold increase in the peak time constant. Without γ-distortion the alignment offset is reduced to 4 mV and the peak time constant is reduced to 0.5 ms, as shown in Fig. 5-8.

The above description for γ-distortion in potassium channels and for sodium inactivation is consistent with experimental results from a number of published papers. A fast gating current component for the potassium channel of the squid giant axon was found by Gilly and Armstrong (1980). Their paper shows the fast potassium gating current component as ending, while the ionic current was just beginning. The gating current

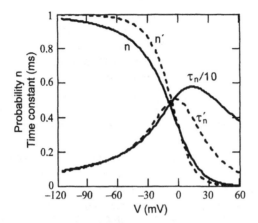

Fig. 5-8. The potassium ion channel gating curves without γ-distortion (dashed lines) are similar to sodium channel curves. The potassium channel curves for n and τ_n were plotted with rate coefficients from Table 6-1B. The curves n' and τ_n' without open-gate distortion, were obtained by setting the γ-distortion factors equal to one. This reduced the peak time constant from 5.8 ms to 0.5 ms and the alignment offset (between V_p and $V_{1/2}$) from 24 mV to 4 mV.

time constant was on the order of 0.5 ms and the ionic current time constant was much longer. Gating current recordings for sodium channel inactivation also show a gating current time constant on the order of 0.5 ms and a much longer time constant for the ionic current (Armstrong and Bezanilla, 1977).

5-6. Edge distortion of inactivation gating

Not all of the distortion for the rate constants could be accounted for with γ-distortion factors. The remaining distortion for sodium inactivation may be due to a nonuniform electric field at the edge of the protein. For the electron-gating model, the equation for α_h should have $N-1$ tunneling sites and be similar to the equation for α_m, but with a voltage polarity reversal. The amino acid sequence for the sodium channel has eight arginine amino acids spaced every third residue in domain IV (Rosenthal and Gilly, 1993). An equation with $N = 8$ was developed that provided a fit to the α_h equation of Hodgkin and Huxley; however, it required substantial correction with calibration coefficients (Table 6-1C).

The need for the large correction is apparently because of the location of the tunneling sites near the inside edge of the protein.

The electron-tunneling rate constants for the inactivation gate have a different sensitivity than the electron-tunneling rate constants for the activation gate. The inactivation rate constants have a sensitivity of 1/11 for β_e and 1/20 for α_e. With the q_8 control site near the protein/cytoplasm interface the electron may experience an electric field, from membrane voltage, with higher flux density, as some of the flux lines converge on one or more charges near the edge, which are not completely neutralized. This would give a higher sensitivity near the edge than near the center of the membrane. However, the 1/20 sensitivity for α_e is less than the 1/18 sensitivity for sodium activation β_e. This suggests that at least some change in the sensitivity may come from another mechanism.

The electron tunneling could be influenced by positive feedback with a nearby positive ion. This interaction could increase the sensitivity for β_e and reduce the sensitivity for α_e. With a small depolarization the electron probability at the q_8 control site increases. This increases the force on an adjacent ion in the cytoplasm and the ion then moves closer to the q_7 and q_8 sites. This, in turn, increases the force on the electron at q_7, which increases its probability to tunnel to q_8 etc. The probability for the electron to be at the q_8 control site is given by

$$P_8 = \frac{1}{1 + \dfrac{\alpha_e}{\beta_e}} = \frac{1}{1 + \dfrac{\exp\left(-\dfrac{V_m + 57}{20}\right)}{\exp\left(\dfrac{V_m + 57}{11}\right)}}, \tag{5.36}$$

and with feedback, a simple representation might be

$$P_8 = \frac{1}{1 + \dfrac{\alpha_e}{\beta_e}} = \frac{1}{1 + \dfrac{\exp\left(-\dfrac{e_0(1-\Delta P)}{e_0}\dfrac{(V_m - V_{1/2})}{2H}\right)}{\exp\left(\dfrac{e_0(1+\Delta P)}{e_0}\dfrac{(V_m - V_{1/2})}{2H}\right)}}. \tag{5.37}$$

The depolarizing ΔV_m causes more charge to move into q_8 than otherwise would occur. This is represented as $e_0(1+\Delta P)$ in Eq. 5.37 and the increase in the charge fraction increases the slope sensitivity for rate constant β_e.

Also, more charge is removed from q_7 than otherwise would occur. This is represented as $e_0(1-\Delta P)$ and it decreases the charge fraction and the slope sensitivity for the rate constant a_e. The undistorted slope sensitivity of $1/2H$ could be near the $1/18$ value for electron tunneling and sodium activation. Expanding this scenario to include all of the tunneling sites might account for the skewing of charge for the α_e rate constant, which results in α_e following a simple exponential instead of the expected curve for seven tunneling sites of a_e.

5-7. Multistate gating

Since the pioneering work of Neher and Sakmann (1976) on patch clamping and single channel recording, it has become accepted that single ion channels generate current pulses, with a constant amplitude, but having a pulse width and frequency with a random characteristic. In the electron-gating model, the random characteristic of the pulses results from thermally activated electron tunneling and the fact that the tunnel-track electrons controlling gating are being controlled by energies below the thermal noise. In a single channel, the response of the tunnel-track electron to membrane voltage and other forces is encoded in the pulse width and frequency of gate openings. Many single channel recordings show a minority of pulses having lesser amplitudes due to subconductance states; often having pulse amplitudes between one-quarter and three-quarters of the maximum amplitude. For the electron-gating model, the energy barrier change required to cause a subconductance state can be calculated from the equation

$$\Delta G_i = -kT \ln A_F , \qquad (5.38)$$

where A_F is the pulse amplitude attenuation fraction and ΔG_i is the gating energy barrier increase above G_0 for a subconductance state i. The energy barrier change depends on the Coulomb force exerted by the tunnel-track electron and its angle with the direction of motion of the ion in crossing the energy barrier. For a subconductance state with a pulse amplitude of one-half the maximum value ($A_F = \frac{1}{2}$), the calculated increase in the barrier from the open-gate value G_0 is $\Delta G_1 = 16.6$ meV. This is about 9% of the 180 meV closed-gate energy barrier change.

For an electron at the q_2 location, next to the control site, the increase in distance from the ion could reduce the Coulomb force to about one-half and a change in the angle of the force could provide another reduction by one-fifth making the q_2 site a possible location for causing this subconductance state. Another possibility for a sodium channel subconductance state is that the electron in the tunnel track IV may produce some effect on activation gating in addition to inactivation gating. The sodium channel q_1 site for domain IV has a cross-sectional angle of $\phi = -60$ degrees away from closest alignment with the pore (Fig. 8-1). It could be causing a subconductance state by slightly closing the channel. By analyzing the frequency of occurrence at different membrane voltages, it may be possible to determine which sites are causing

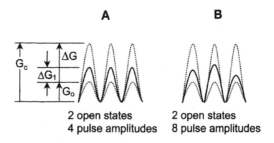

A   **B**

2 open states   2 open states
4 pulse amplitudes   8 pulse amplitudes

Fig. 5-9. Possible multistate gating for the sodium channel. The open gate energy barrier G_o is taken as the reference. Fig. *A* shows three activation energy barriers, each with an identical subconductance state ΔG_1. Combinations of the two open states (G_o and $G_o + \Delta G_1$) give four current pulse amplitudes. If each energy barrier has a different subconductance state ΔG_n, as in Fig. *B*, then eight different current pulse amplitudes result.

subconductance pulses. Figure 3-3 indicates how the electron probability peaks at different voltages for a 4-site tunnel track. Subconductance pulses having a very long duration are likely to be due to *far sites* or to *back sites* (described in Chapter 8) and their average duration should correlate with the time constant determined by the tunneling distance. Multiple subconductance states might be accounted for by different alignments in the q_2 location of tunnel tracks I through III. Fig. 5-9 illustrates how up to eight pulse amplitudes could result from one subconductance state with a different energy barrier amplitude for each of the three sodium activation energy barriers.

Chapter **6**

CHARACTERIZATION AND VALIDATION

One objective was to link the electron-gating model to experimental data wherever possible and to account for any differences between the model and the experimental data. The principal links to experimental data are the following:

1. A primary experimental link is the Hodgkin-Huxley equations. A detailed study was made to understand the significance of the coefficients and sensitivity factors in terms of the electron-gating model. This is described in Chapter 5.
2. The amino acid sequence data for the squid giant axon for sodium and potassium ion channels and the sequence data for *Shaker* B (Figs. 8-1, 8-2 and Figs. 9-1, 9-2).
3. The experimental data for gating current and gating current time constants as described in Section 8-5.
4. Experimental data for electron tunneling across proteins and the α-helix in particular.
5. Experimental determination of the NH_3 inversion frequencies, as described in Part II. The measured frequencies narrow the range of possible parameter values.

6-1. Electron-gating model equations

Equations for the electron-gating model, shown in Table 6-1*B* and *C*, were

Table 6-1. Sodium and potassium ion channel equations

Hodgkin-Huxley A) equations	Electron-gating model equations B) Using displacement voltage (V)
Sodium activation	$N = 4 \quad H = 9 \quad c_2 = 0.68 \quad c_3 = 1.08 \quad c_4 = 0.9$

$$\alpha_m = \frac{0.1(V+25)}{\exp\left(\dfrac{V+25}{10}\right)-1} \qquad\qquad \alpha_m = \frac{1}{\dfrac{1}{N-1}\displaystyle\sum_{i=2}^{N}\exp\left(\dfrac{2i-3}{N-1}c_i\dfrac{V+25}{2H}\right)}$$

$$\beta_m = 4\exp\left(\frac{V}{18}\right) \qquad\qquad \beta_m = \exp\left(\frac{V+25}{2H}\right)$$

Sodium inactivation $\qquad H = 5.5 \quad k'T = 24\text{ mV} \quad E_r = -60\text{ mV} \quad E_{Na} = 55\text{ mV}$
$\qquad\qquad\qquad\qquad\qquad \Delta G_h = 180\text{ meV} \quad s = 0.55$

$$\alpha_h = 0.07\exp\left(\frac{V}{20}\right) \qquad\qquad \alpha_h = \exp\left(s\frac{V+3}{2H}\right)\gamma_{\alpha h} \qquad \gamma_{\alpha h} = b_h = 0.060$$

$$\beta_h = \frac{1}{\exp\left(\dfrac{V+30}{10}\right)+1} \qquad\qquad \beta_h = \exp\left(-\frac{V+2}{2H}\right)\gamma_{\beta h}$$

$$\gamma_{\beta h} = \frac{b_h\exp\left(-\dfrac{V}{4.58k'T}\right)}{1+\exp\left(\dfrac{-\Delta G_h}{kT}+\dfrac{E_{Na}-E_r}{k'T}-\dfrac{V+3}{2H}-\dfrac{V}{4.58k'T}\right)}$$

Potassium $\qquad\qquad\qquad\qquad N = 7 \quad H = 13 \quad c_2 = 1.53 \quad c_3 = 2.4 \quad c_4 = 1.7$
$\qquad\qquad\qquad\qquad\qquad\qquad c_5 = 1.2 \quad c_6 = 1.2 \quad c_7 = 1.5 \quad k'T = 24$

$$\alpha_n = \frac{0.01(V+10)}{\exp\left(\dfrac{V+10}{10}\right)-1} \qquad\qquad \alpha_n = \frac{\gamma_{\alpha n}}{\dfrac{1}{N-1}\displaystyle\sum_{i=2}^{N}\exp\left(\dfrac{2i-3}{N-1}c_i\dfrac{V}{2H}\right)}$$

$$\beta_n = 0.125\exp\left(\frac{V}{80}\right) \qquad\qquad \beta_n = \exp\left(\frac{V}{2H}\right)\gamma_{\beta n}$$

$$\gamma_{\beta n} = b_n\exp\left(-\frac{V-29}{1.6k'T}\right) \qquad \gamma_{\alpha n} = b_n = 0.059$$

Table 6-1.

	Electron-gating model equations
	C) Using amplification (h_w) and membrane voltage (V_m)

Sodium activation

$N = 4$
$h_w = 25.2$
$\eta = 0.106$
$c_2 = 0.68$
$c_3 = 1.08$
$c_4 = 0.9$
$k'T = 24$ mV
$V_{1/2} = -35$ mV

$$\alpha_m = \cfrac{1}{\cfrac{1}{N-1}\sum_{i=2}^{N}\exp\left(-h_w\eta\,\frac{2i-3}{N-1}c_i\,\frac{V_m-V_{1/2}}{2k'T}\right)}$$

$$\beta_m = \exp\left(-h_w\eta\,\frac{V_m-V_{1/2}}{2k'T}\right)$$

Sodium inactivation

$N = 8$
$h_w = 25.2$
$\eta = 0.173$
$s = 0.55$
$c_2 = 3.9 \quad c_6 = 0.8$
$c_3 = 2.4 \quad c_7 = 0.5$
$c_4 = 1.2 \quad c_8 = 0.5$
$c_5 = 1$
$\gamma_{\alpha h} = b_h = 0.060$
$\Delta G_h = 180$ meV
$kT = 24$ meV
$k'T = 24$ mV
$E_{Na} = 55$ mV
$V_{1/2} = -57$ mV

$$\alpha_h = \cfrac{\gamma_{\alpha h}}{\cfrac{1}{N-1}\sum_{i=2}^{N}\exp\left(sh_w\eta\,\frac{2i-3}{N-1}c_i\,\frac{V_m-V_{1/2}}{2k'T}\right)}$$

$$\beta_h = \exp\left(h_w\eta\,\frac{V_m-V_{1/2}}{2k'T}\right)\gamma_{\beta h}$$

$$\gamma_{\beta h} = \cfrac{b_h\exp\left(\cfrac{V_m+70}{4.58k'T}\right)}{1+\exp\left(\cfrac{-\Delta G_h}{kT}+\cfrac{E_{Na}-V_{1/2}}{k'T}+h_w\eta\,\cfrac{V_m-V_{1/2}}{2k'T}+\cfrac{V_m+73}{4.58k'T}\right)}$$

Potassium

$N = 7$
$h_w = 25.2$
$\eta = 0.074$
$c_2 = 1.53 \quad c_5 = 1.2$
$c_3 = 2.4 \quad c_6 = 1.2$
$c_4 = 1.7 \quad c_7 = 1.5$
$\gamma_{\alpha n} = b_n = 0.059$
$E_K = -75$ mV
$k'T = 24$ mV
$V_{1/2} = -48$ mV

$$\alpha_n = \cfrac{\gamma_{\alpha n}}{\cfrac{1}{N-1}\sum_{i=2}^{N}\exp\left(-h_w\eta\,\frac{2i-3}{N-1}c_i\,\frac{V_m-V_{1/2}+12}{2k'T}\right)}$$

$$\beta_n = \exp\left(-h_w\eta\,\frac{V_m-V_{1/2}+12}{2k'T}\right)\gamma_{\beta n}$$

$$\gamma_{\beta n} = b_n\exp\left(\frac{V_m-E_K+14}{1.6k'T}\right)$$

developed in Chapters 3 and 5. The calibration coefficients, labeled as indexed c_i terms, were selected for a close match of the α rate curves with the curves for the Hodgkin-Huxley equations. They compensate for distortions, caused by local charges that create a non-uniform force field across the tunnel tracks. The c_i terms separately multiply the slope sensitivity for each of the summed exponential terms. Another type of compensation using an indexed offset voltage was tried, but this method did not provide satisfactory curve matching. This suggested that the distorting force field, acting on the electron at each tunneling site was voltage dependent, since a fixed field would be best compensated by the index offset method. The correction coefficients have the greatest deviation from unity in the two positions closest to the control site. From this it appears that the force field may be partly due to the positively charged ions transiting through the gating sites. The ions could increase the probability for the electron to be at the q_2 or q_3 tunneling sites and increase voltage dependence of the rate constant by a regenerative process. This could require an increase in the value of coefficient c_1 and/or c_2 for curve matching. A particularly large correction factor is required for coefficient c_1 and c_2 in the equation for α_h. This is likely due to the inactivation gate's location near the protein/cytoplasm interface. A regenerative process that could account for an increase in the sensitivity might be as follows:

A positive ion in the cytoplasm is near an electron at the q_7 site. The force between the ion and the electron increases the probability for the electron to tunnel to the q_8 control site, which in turn reduces the mean distance to the mobile ion, which further increases the force and the electron probability at the q_8 site. This regenerative effect, requiring a large increase in coefficient c_1, may distort the curve to approximately that of a simple exponential, as shown for α_h in Table 6-1A and B.

The equation for α_h in Table 6-1C provides an alternate $N = 8$ model to match the eight arginine/lysine sites in the amino acid sequence. To be consistent, all tunnel tracks for a channel are labeled with sites q_1 through q_n increasing in one direction, as described in Section 8-1. Because the inactivation control site is at q_8, the index i (for α_h) increments in the reverse direction, going from two at site q_7 to eight at site q_1.

The $\gamma_{\beta h}$ distortion factor for β_h is for a simplification of Eq. 5.8. The exponent a' is incorporated into the equation, so that for a displacement voltage more negative than -100 mV ($V_m > 40$ mV), the voltage terms in the numerator and denominator of β_h will always cancel. This produces a constant rate in this region of the curve, even for a different value of H or a different value for η.

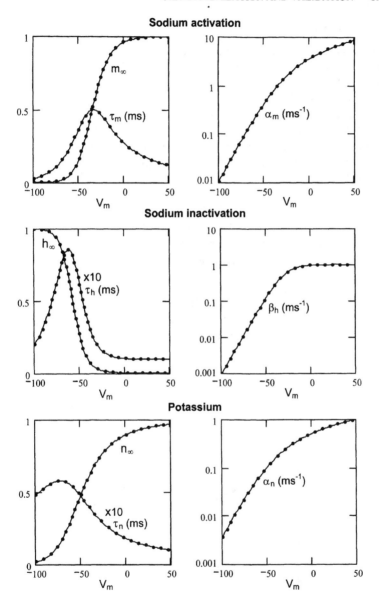

Fig. 6-1. A comparison showing the close match of the electron-gating model curves (solid line) with corresponding curves for the Hodgkin-Huxley equations (dotted line). The Hodgkin-Huxley equations were converted to membrane voltage by replacing V with $E_r - V_m$. The electron gating model curves use the rate coefficients and parameters from Table 6-1C.

6-2. Finite-range rate constants

According to the model, ion channel rate constants can be written in two forms: a simple form having unlimited range (Table 6-1) and a more complicated form having a finite range (described in Section 4-1). The latter form results from combining the former with the Marcus equation for the inverted region of electron tunneling. Starting with Eq. 4.7, the squared energy terms in the numerator can be expanded. At low free energy, the term $e_0 v$ squared can be neglected and the fixed λ term separated out; then, multiplying by $K_0 h'_w / K_w$ and simplifying the expression gives a rate constant for electron tunneling.

$$\beta_e = \frac{K_0 h'_w \exp\left(\dfrac{-v}{2k'T}\right)}{1 + \dfrac{h'_w\left(h'_w - 1\right)}{1 + \left(h'_w - 1\right)\exp\left[\dfrac{-\left(h'_w - 1\right)v}{2k'T}\right]}} \tag{6.1}$$

Each of the amplifying NH_3 electron tunneling sites across the protein can be represented by Eq. 6.1 for β_e (Fig. 6-2) or with opposite polarity as α_e. The corresponding equation for β having an unlimited range is given by

$$\beta = K_0 \exp\left(\frac{-h_w v}{2k'T}\right), \tag{6.2}$$

and is shown as the dashed line tangent to β_e.

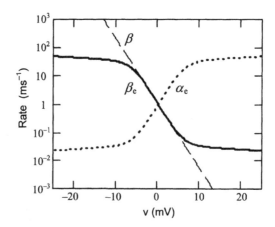

Fig. 6-2. Amplified tunneling rate versus tunneling voltage

To have the same slope sensitivity, the amplification factor h'_w needs to be larger than h_w. The value of $h'_w = 28$ was used instead of $h_w = 25.2$.

Because the finite-range equations have a large number of terms, it is convenient to have a standardized pair of equations for representing the sodium and potassium rate constants and a table for assigning parameter values. Table 6-2 gives the finite-range equations for the electron-gating model that correspond to the equations in Table 6-1C. The curves for these equations were plotted against the displacement voltage ($E_r - V_m$) to facilitate a comparison with the graphs and data of Hodgkin and Huxley (1952). The results are shown in Fig. 6-3.

The distortion factors are the same as developed in Chapter 5, except for the factor $\gamma_{\beta h}$. There was a problem obtaining a graph for β_h, matching the Hodgkin-Huxley curve. The distortion factor $\gamma_{\beta h}$ includes the ratio β_e/α_e. Both β_e and α_e saturate (Fig. 6-2). For a displacement voltage in the -100 mV ($V_m = 40$ mV) region, the saturating β_e term (Eq. 6.9) cancels with β_e for Eq. 6.4. The α_e term does not cancel with anything and the saturation distorts the curve, preventing a match up with the β_h reference curve. The solution was to include the distortion factor γ_{oh}. In developing the leakage distortion factor γ_{ch} (Section 5.1), the energy barrier in the close state (ΔG_h) represents the increase in barrier height above the open-gate barrier. The open-gate energy barrier G_{oh} was taken as a constant reference; but the effective height of the barrier varies with membrane voltage, as given by γ_{oh}. The factor γ_{oh}, representing modulation of the open-gate energy barrier, needs to be included in the equation for the closed-gate leakage distortion, as in Eq. 6.8.

The γ_{oh} factor compensates for the reduction in slope (saturation) near the end of the amplification range and facilitates an acceptable match-up with Hodgkin-Huxley data (Fig. 6-3). Interestingly, this factor cannot be included in the unlimited range equation for β_h in Table 6-1 and still have agreement with the reference curve. This seems to indicate that saturation is required for agreement with the Hodgkin-Huxley equation for inactivation and it lends some measure of experimental support for the finite-range rate constants.

A comparison of the finite-range rate constants with the Hodgkin-Huxley rate constants is shown in Fig. 6-3, covering the range of recorded ion channel data. Just beyond the range of the Hodgkin-Huxley data, the finite-range rate constants enter the non-amplifying, saturating region. That these rate curves are in reasonable agreement with the Hodgkin-Huxley data over the range of recorded data, and not further, might be taken as a further indication of the validity of the finite-range rate constants.

Unified, finite-range sodium and potassium ion channel rate constants

$$\alpha_x = \frac{K_0 \gamma_{\alpha x}}{\left[\dfrac{1}{N-1} \displaystyle\sum_{i=2}^{N} 1 + \dfrac{k_z h_w' \exp\left(as\eta \dfrac{V_m - V' - V_z}{2k'T} \right)}{1 + \left(h_w' - 1 \right) \exp\left(as\eta \left(h_w' - 1 \right) \dfrac{V_m - V' - V_z}{2k'T} \right)} \right]^{\left(\frac{2i-3}{N-1} \right) c_i}} \quad (6.3)$$

$$\beta_x = \frac{k_z h_w' \exp\left(a\eta \dfrac{V_m - V' - V_z}{2k'T} \right) K_0 \gamma_{\beta x}}{1 + \dfrac{h_w' \left(h_w' - 1 \right)}{1 + \left(h_w' - 1 \right) \exp\left(a\eta \left(h_w' - 1 \right) \dfrac{V_m - V' - V_z}{2k'T} \right)}} \quad (6.4)$$

Table 6-2.

Sodium activation	Potassium	Sodium inactivation
$x = m$	$x = n$	$x = h$
$N = 4$	$N = 7$	$N = 8$
$h_w' = 28$	$h_w' = 28$	$h_w' = 28$
$\eta = 0.106$	$\eta = 0.074$	$\eta = 0.173$
$a = -1$	$a = -1$	$a = 1$
$s = 1$	$s = 1$	$s = 0.55$
$c_2 = 1.35$	$c_2 = 2.32$	$c_2 = 5.9$
$c_3 = 0.54$	$c_3 = 1.3$	$c_3 = 2.4$
$c_4 = 0.95$	$c_4 = 1.7$	$c_4 = 1.3$
$V' = -35$ mV	$c_5 = 1.6$	$c_5 = 1.0$
$V_z = 0$ mV	$c_6 = 1.2$	$c_6 = 0.9$
$k_z = 1$	$c_7 = 1.5$	$c_7 = 0.6$
$k'T = 24$ mV	$V' = -60$ mV	$c_8 = 0.5$
$\gamma_{\alpha m} = 1$	$V_z = 12$ mV	$V' = -56$ mV
$\gamma_{\beta m} = 1$	$k_z = 0.629$	$V_z = 0$ mV
$K_0 = 1$ ms^{-1}	$E_K = -75$ mV	$k_z = 1$
	$k'T = 24$ mV	$E_{Na} = 55$ mV
	$\gamma_{\alpha n} = b_n = 0.059$	$G_{oh} = 50$ meV
	$\gamma_{\beta n} = $ Eq. 6.5	$N_F = 1.02$
	$K_0 = 1$ ms^{-1}	$\Delta G_h = 190$ meV
		$k'T = 24$ mV
		$a' = 0.90$
		$\gamma_{\alpha h} = b_h = 0.060$
		$\gamma_{\beta h} = $ Eq. 6.8

Distortion factors

The following equations are for γ-distortion factors. They are computed before computing rate constants with Eqs. 6.3 and 6.4.

Potassium

$$\gamma_{\beta n}= b_n \exp\left(\frac{V_m-E_K+14}{1.6k'T}\right) \tag{6.5}$$

Sodium Inactivation

$$\alpha_e=\frac{\alpha_h}{\gamma_{\alpha h}} \qquad \beta_e=\frac{\beta_h}{\gamma_{\beta h}} \tag{6.6}$$

$$\gamma_{oh}=\frac{N_F}{1+\exp\left(\dfrac{G_{oh}}{kT}+\dfrac{V_m-E_{Na}}{k'T}\right)} \tag{6.7}$$

$$\gamma_{\beta h}=\frac{b_h\exp\left(\dfrac{V_m+70}{4.58k'T}\right)}{1+\exp\left(\dfrac{-\Delta G_h}{kT}+\dfrac{E_{Na}-V_m}{k'T}\right)\left(\dfrac{\beta_e}{\alpha_e}\right)^{a'}\dfrac{1}{\gamma_{oh}}\exp\left(\dfrac{V_m+70}{4.58k'T}\right)} \tag{6.8}$$

In some cases a voltage shift V_z may be needed to center the operating membrane voltage range within the amplifying region (Fig. 6-2). A factor k_z can be calculated to restore the correct calibration after the V_z shift.

$$k_z = \frac{1}{h'_w}+\frac{h'_w-1}{1+\left(h'_w-1\right)\exp\left[-asn\left(h'_w-1\right)\dfrac{V_z}{2k'T}\right]} \tag{6.9}$$

There is a $V_z= 12$ mV shift for the potassium channel α and β rate constants associated with the γ-distortion factor. It was removed, since it is for ion transport and should not affect centering within the amplifying voltage range. The 12 mV shift was then restored by multiplying by $k_z = 0.629$. Without centering there was distortion of the rate constants near the end of the voltage range, due to saturation.

Note: The equations in this section are superseded by the rate equations of Table 6-1C using the voltage-sensitive amplification factor $h_w(V_m)$ (Eqs. 2.39 and 2.41). The equations and graphs are shown for comparison.

**Fig. 6-3. Finite-range rate constants (solid line) versus Hodgkin-Huxley
rate constants**

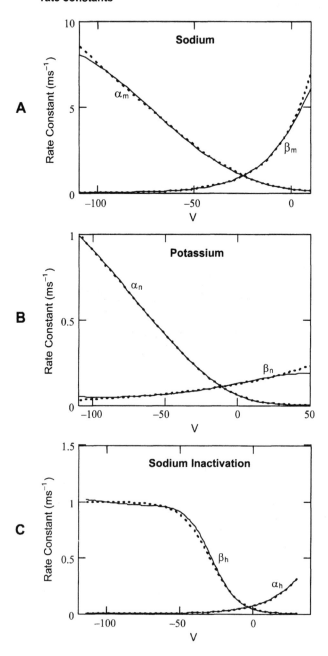

6-3. Open-channel probability range and time constant

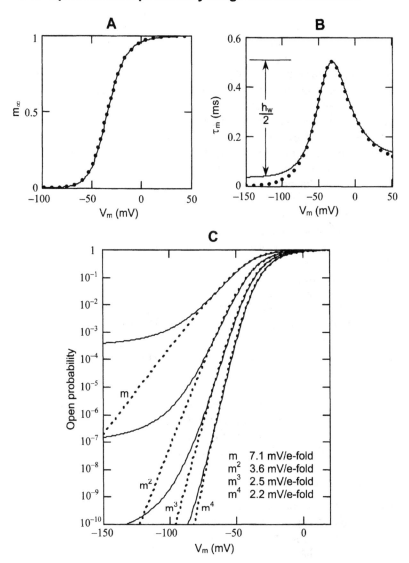

Fig. 6-4. Characteristic curves for the sodium channel are calculated with the finite-range rate constants α_m and β_m (Eqs. 6.3 and 6.4). Parameter values are from Table 6-2, except for different values of $V_z = 4$ mV and $K_z = 0.80$. The dotted lines were calculated with the Hodgkin-Huxley rate constants.

The gating curves in Fig. 6-4 are for sodium channel activation using electron gating and tunnel-tracks that have four arginine sites ($N = 4$). The dotted lines are for the Hodgkin-Huxley equations. In the electron-gating model, the finite-range time-constant curve (B) saturates with a hyperpolarizing voltage, limiting the time-constant range to $(h_w-1)/2$. With depolarization the range is even less, because the electron probability is distributed across the $N-1$ sites of the α-rate constant.

This peak-to-valley ratio, given by $(h_w-1)/2$ is a useful factor for comparing with experimental observations, since it does not depend on the electric field or the attenuation factor, but it does depend on the tunneling distance. For electrons tunneling to far sites, h_w decreases as the tunneling distance increases, as indicated in Table 8-1.

The open-channel probability has a range (before saturation) that depends upon the number of tunnel-tracks used for activation. With the standard m^3 model for the sodium channel, the curve fits an exponential down to 10^{-6}. A study of noninactivating sodium channels has shown that the open probability fits an exponential down to 10^{-7}, with a limiting slope of 2.2 mV per e-fold (Hirschberg et al., 1995; see also Hille, 2001).

The easiest way for the electron-gating model to have agreement with this data would be to increase the exponent to m^4. This extends the range to beyond 10^{-7} and gives a slope of 2.2 mV per e-fold. It seems reasonable that the sodium channel domain IV electron might contribute to activation gating. The q_1 site for tunnel track IV is likely in a staggered alignment with the other three q_1 sites. The gating energy barrier would depend on the residues in the adjacent cavity of domain IV. An interesting comparison is with potassium channels, which have both activation and inactivation gates using the same tunnel-track (Chapter 9). Going from an m^3 to an m^4 model for the sodium channel should require relative few changes. The main effect would be to cause a 4 mV rightward shift in the midpoint voltage for the m^4 curve.

6-4. Rate curves using voltage-sensitive amplification

Saturation for the rate constants is needed to allow a merger with the non-amplifying higher free energy region that can occur with electron tunneling (Fig. 4-2). Another way to have saturation and finite-range rate constants is to have voltage-sensitive amplification. It was discovered that the dipole moment of NH_3, interacting with the donor electron, would make amplification voltage-sensitive. Using $h_w(V_m)$, the curves of Fig. 6-3 and Fig. 6-4 were re-plotted to check for agreement with Hodgkin-Huxley data. Curves were plotted using Table 6-1 equations and Eqs. 2.39 and 2.41.

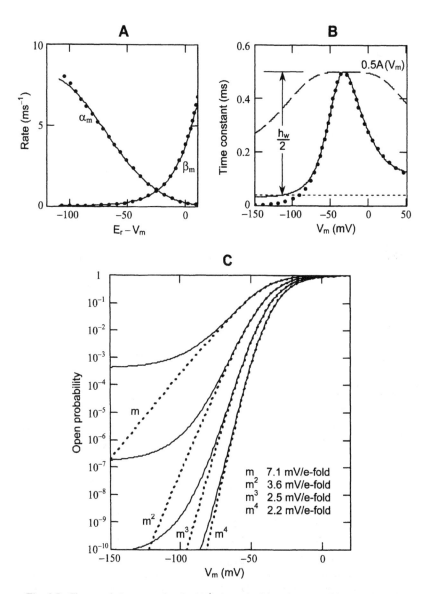

Fig. 6-5. Characteristic curves for the Na⁺ channel with voltage sensitive amplification (solid lines) are compared with curves for the Hodgkin-Huxley rate constants (dotted lines). The variation of amplification with voltage, indicated by $A(V_m)$, causes saturation of the β rate constant curve (Fig. 2-3B) and a plateau in the time constant curve (B). The curves are in agreement with curves using the finite-range rate constants (Eqs. 6.3 and 6.4).

Table 6-3.

Sodium activation	Potassium	Sodium inactivation
$N = 4$	$N = 7$	$N = 8$
$h_w = 25.2$	$h_w = 25.2$	$h_w = 25.2$
$\eta = 0.106$	$\eta = 0.074$	$\eta = 0.173$
$c_2 = 1.23$	$c_2 = 2.07$	$c_2 = 3.9$
$c_3 = 0.6$	$c_3 = 1.6$	$c_3 = 2.0$
$c_4 = 0.95$	$c_4 = 1.7$	$c_4 = 1.2$
$V_0 = -35$ mV	$c_5 = 1.5$	$c_5 = 1.0$
$V_{1/2} = -35$ mV	$c_6 = 1.2$	$c_6 = 0.8$
$k'T = 24$ mV	$c_7 = 1.5$	$c_7 = 0.7$
$K_0 = 1$ ms^{-1}	$V_0 = -48$ mV	$c_8 = 0.5$
$E_r = -60$ mV	$V_{1/2} = -48$ mV	$V_0 = -57$ mV
	$E_K = -75$ mV	$V_{1/2} = -57$ mV
	$k'T = 24$ mV	$E_{Na} = 55$ mV
	$\gamma_{\alpha n} = b_n = 0.059$	$\Delta G_h = 186$ meV
	$K_0 = 1$ ms^{-1}	$k'T = 24$ mV
		$\gamma_{\alpha h} = b_h = 0.060$
		$K_0 = 1$ ms^{-1}

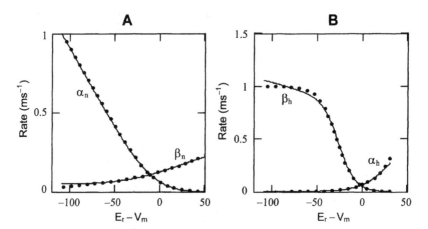

Fig. 6-6. Curves for the potassium channel and for sodium-inactivation rate constants with voltage-sensitive amplification (solid lines) are compared with curves for the Hodgkin-Huxley rate constants (dotted line). The voltage sensitive amplification factor from Eqs. 2.39 and 2.41 were used in conjunction with the equations of Table 6-1C. Parameter values are shown in Table 6-3. The curves were plotted covering the same displacement voltage range as for the Hodgkin-Huxley data.

Chapter 7

FLUX GATING IN Na⁺ AND K⁺ CHANNELS

7-1. Sodium channel flux gating

In the previous chapter, the electron-gating model rate coefficients were compared with the Hodgkin-Huxley equations. These equations help to define a linear model, where channel current is given by the equation $I_{Na} = g_{Na} m^3 h (V_m - E_{Na})$ or by $I_K = g_K n^4 (V_m - E_K)$. It seems appropriate here to compare the ion channel current for the linear model with the channel current computed as the difference between the influx and the efflux (or vice-versa) using the Goldman, Hodgkin and Katz relation in conjunction with gating. The I-V curve in Fig. 1-3, for the sodium channel, illustrates the difference in the two methods. The GHK flux difference gives the current for an ungated or fully open channel with low energy barriers. To represent gating, the fluxes are multiplied by the open channel probability. Studies have shown that the GHK flux difference model, for the sodium channel, provides good agreement with experimental data even though it does not have any specific energy barrier terms.

In Section 5-4, it was determined that there was a negligible amount of γ-distortion in sodium channel activation and therefore, m^3 can be taken as the open-channel probability (P_o), if a value of one is assigned to h. The results, plotted in Fig. 7-1, indicate agreement between the two methods of computing current up to a membrane potential of $-10\,\text{mV}$. At a more positive potential there is a divergence, from the linear model curve (I_{Na}), which has a slower decline in current as the membrane is

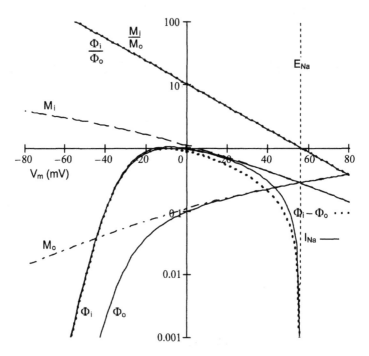

Fig. 7-1. Sodium channel gated and ungated flux curves are shown along with the linear model current (I_{Na}) curve and the gated flux difference. The curve for Na^+ ion current, in the linear model, is in agreement with the gated flux difference up to -10 mV. As the channel opens further, the Na^+ current computed as the gated flux difference, decreases more rapidly. The curves for I_{Na} and flux difference were scaled to match the gated influx curve.

further depolarized. The gated flux ratio (Φ_i/Φ_o) and the ungated GHK flux ratio (M_i/M_o) have the same value over the entire voltage range.

The α_m and β_m terms, shown in the calculations, are for the Hodgkin-Huxley rate coefficients converted to membrane potential. A shift of 1 mV (Eq. 7.4) improved the alignment of I_{Na} with the gated flux difference curve.

7-2. Sodium channel inactivation flux gating

Unlike sodium channel activation, inactivation does have γ-distortion due to modulation of open-channel energy barriers by flux-ratio parameters of the Ussing equation. This distortion is described in Section 5-3. The linear model current I_{Na} includes this distortion in the open-gate probability term h. For computing gated fluxes, the term P_o is used to refer to the open channel probability without this distortion. By removing this modulation from the β_h rate constant, the slope sensitivity is reduced from 1/10 to 1/11. This slightly reduced sensitivity is used for β_{eh} in Eq. 7.6. Rate constant β_{eh} is for electron tunneling, away from the inactivation-gate control site q_8.

In Fig. 7-2, the linear model current I_{Na} is in agreement with the gated flux difference over the full operating range of membrane potential. The agreement provides a confirmation of the reduced slope sensitivity for β_{eh}. The gentle slope of the current curves is partially the result of leakage over a single closed-gate energy barrier. Interestingly, the gated and ungated flux ratios are (almost) identical for sodium channel inactivation and thus the Ussing flux-ratio test must be obeyed.

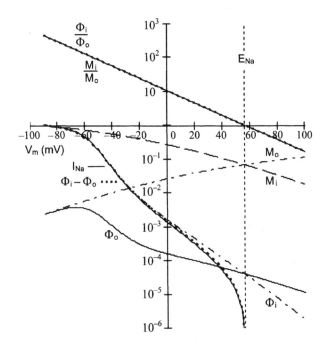

Fig. 7-2. Sodium channel inactivation gated and ungated flux curves are shown along with the linear model current (I_{Na}) curve and the gated flux difference. The curves show agreement between the two methods of computing current, over the normal operating range of membrane potential. The bowed shape of the gated influx and efflux curves is from the closed-gate leakage characteristic of rate constant β_h.

Equations used in plotting the sodium channel curves

GHK equations:

$$M_i = \frac{K'[S]_o}{\exp\left(\dfrac{V_m}{k'T}\right) - 1} V_m \qquad M_o = \frac{K'[S]_i \exp\left(\dfrac{V_m}{k'T}\right)}{\exp\left(\dfrac{V_m}{k'T}\right) - 1} V_m \tag{7.1}$$

Sodium ion concentration: (Temp. 6°C)

$$[S]_o = 500\,\text{mM} \quad [S]_i = 50\,\text{mM} \quad E_{Na} = 24\ln\left(\frac{[S]_o}{[S]_i}\right) = 55.2\,\text{mV} \tag{7.2}$$

Flux gating equations:

$$\Phi_i = P_o M_i \qquad \Phi_o = P_o M_o \tag{7.3}$$

For sodium activation: GHK calibration: $K' = 9.0 \times 10^{-5}$

$$P_o = \left(\frac{1}{1 + \dfrac{\beta_m(V_m)}{\alpha_m(V_m)}}\right)^3 \qquad m^3 = \left(\frac{1}{1 + \dfrac{\beta_m(V_m + 1)}{\alpha_m(V_m + 1)}}\right)^3 \tag{7.4}$$

Linear model I_{Na}:
$$I_{Na} = m^3(V_m - E_{Na})\left(\frac{-1.04}{E_{Na}}\right) \tag{7.5}$$

For sodium inactivation: GHK calibration: $K' = 2.15 \times 10^{-5}$

$$P_o = \frac{1}{1 + \dfrac{\beta_{eh}}{\alpha_h}} \qquad \beta_{eh} = \exp\left(\frac{V_m + 60}{11}\right) \frac{0.05}{1 + \exp\left(\dfrac{V_m + 30}{10}\right)} \tag{7.6}$$

$$h = \frac{1}{1 + \dfrac{\beta_h}{\alpha_h}} \qquad \alpha_h = 0.07\exp\left(-\frac{V_m + 60}{20}\right) \qquad \beta_h = \left[1 + \exp\left(-\frac{V_m + 30}{10}\right)\right]^{-1} \tag{7.7}$$

Linear model I_{Na}:
$$I_{Na} = h(V_m - E_{Na})\left(\frac{-0.36}{E_{Na}}\right) \tag{7.8}$$

7-3. Potassium channel flux gating

Unlike sodium channels, potassium channels do not obey the Ussing (1949) flux ratio test. The Ussing flux ratio gives a relationship between the efflux and influx for passive, independent transport of ions across a membrane. For potassium channels it has the form:

$$\frac{M_o}{M_i} = \frac{[K]_i}{[K]_o}\exp\left(\frac{V_m}{k'T}\right) = \exp\left(\frac{V_m - E_K}{k'T}\right), \qquad (7.9)$$

where k' is equal to k/e_0 or R/F. Hodgkin and Keynes (1955) discovered that the predictions of the flux ratio test for potassium channels of *Sepia* giant axons were not obeyed. Their observations were better described with the equation raised to an exponent of $n' = 2.5$. This result was confirmed by Begenisich and De Weer (1980) for potassium channels of the squid giant axons. They found that the value for n' varied with membrane voltage, but was generally between 2 and 3. In their paper, Hodgkin and Keynes (1955) developed a knock on theory to explain the exponent n' being greater than one. Flux ratio exponents greater than one have generally been interpreted as an interaction of multiple ions, simultaneously crossing the narrow region of the channel.

To better understand what is happening in the potassium channel, a flux-gating model was developed. The product of open-channel probability (P_o) with the influx and efflux of the Goldman-Hodgkin-Katz equations was the starting point. Then, the distortions of the influx and efflux needed to be, at least approximately, described. A set of criteria was compiled that the model should meet, which included the following:

1. The gated flux ratio must always be equal to one when V_m equals E_K. This includes any variations in the gating probability curve.
2. As the channel opens fully, the gated influx and efflux curves should merge with the curves for the GHK fluxes.
3. The model should be in agreement with experimental observations.

It was assumed that without gating, potassium channels could be described, at least approximately, by the GHK relation, and that flux gating results in the distortions. Thus, when the channel fully opens the gated fluxes merge with the ungated fluxes, as shown in Fig. 7-3.

Calibrating the gated influx and efflux curves
Curves for the gated fluxes were plotted (Fig. 7-3) using Eq. 7.11 and

Eq. 7.12, which were developed to meet the above criteria. Equations 7.14 and 7.15 are for distortion factors that help match the efflux and especially the influx to experimental data. The most striking characteristic is the large attenuation for the gated influx. The gated influx curve Φ_i was adjusted to provide approximate agreement with Fig. 6 in the Hodgkin and Keynes (1955) paper. The efflux curve Φ_o is also in approximate agreement with their curve, except for a steeper slope. An interesting characteristic of their graph is the upward slope of the influx as it crosses the efflux. This portion of the curve is well described by Eq. 7.14, which represents closed-gate leakage (L_i) and has a similar form to the equation for sodium channel inactivation gate leakage. The upward slope can be adjusted with the energy barrier ΔG_n or sensitivity factor 1/9. The distortion factor D_o, for the gated efflux, shifts the lower part of the Φ_o curve (Fig. 7-3) in the negative direction, to give approximate agreement with the experimental observations. A ten fold increase in external potassium concentration (with $V_m = -30$ mV) reduces the efflux by a factor of 2.4, about the same as the 2.6 value reported by Hodgkin and Keynes. The efflux distortion factor D_o, given by Eq. 7.15, is responsible for this.

Additional calibration of the graph was made by adjusting amplitude parameter A_1 (Eq. 7.12) for a flux ratio exponent (Eq. 7.17) equal to $n' = 3$ at $V_m = -30$ mV. This gave a value for n' in reasonable agreement with the observations of Begenisich and De Weer (1980), as shown in Fig. 7-4. Their weighted least squares regression line (Fig. 3 in their paper) is the dotted line in Fig. 7-4.

Having a flux ratio always equal to one when $V_m = E_K$ was an elusive problem at first, but this was finally solved with the ratio $P_o(V_m)/P_o(E_K)$ in the denominator of the influx equation. For the $P_o(E_K)$ term, open channel probability P_o is evaluated with E_K substituted for V_m. This defines a reference point for the influx, and when $V_m = E_K$ the denominator terms cancel and the other terms in the influx and efflux also cancel, leaving a flux ratio of one. It was found that the influx curve was best described with $P_o(V_m)$ and $P_o(E_K)$ raised to an exponent of 3/4. This is thought to be a result of the influx gating latch-up effect, described in Section 7-4. With the denominator much greater than one, the term $[P_o]^{3/4}$ partially cancels the open channel probability P_o in the numerator, leaving the influx dependent on $[P_o]^{1/4}$. This dependency suggests the involvement of only one of the four gated energy barriers in the influx attenuation.

Calculating the flux ratio exponent n'

The flux ratio exponent n' was determined by the deviation of the gated flux ratio from the ungated flux ratio (Fig. 7-3). Values for n' were

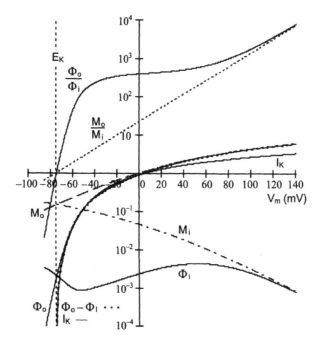

Fig. 7-3. Potassium channel gated and ungated flux curves are shown, along with the linear model current I_K and the gated flux difference. Gating of the GHK fluxes M_i and M_o, results in a much larger attenuation for the influx Φ_i than for the efflux Φ_o. This causes Φ_o/Φ_i to be greater than M_o/M_i and the flux ratio exponent n' to be greater than one. The linear model current I_K is in agreement with the gated flux difference up to a membrane potential of 10 mV.

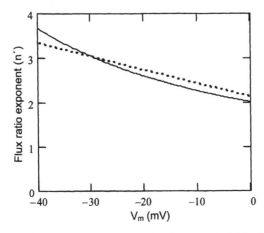

Fig. 7-4. Comparison of flux ratio exponent for the model (solid line) with a weighted least squares regression line for observed values of n' (dotted line).

Equations used in plotting the potassium channel curves

GHK equations:

Same as Eqs. 8-1 with $[K] = [S]$ GHK calibration: $K' = 1.2 \times 10^{-4}$

$[K]_o = 15.4 \, \text{mM}$ $[K]_i = 350 \, \text{mM}$ $E_K = 24 \ln\left(\dfrac{[K]_o}{[K]_i}\right) = -75 \, \text{mV}$ (7.10)

Flux gating equations:

$$\Phi_o = P_o D_o M_o \tag{7.11}$$

$$\Phi_i = \dfrac{P_o D_o M_i}{\left[\dfrac{1 + \exp\left(-\dfrac{V_m + E_K}{k'T}\right)\left[P_o(V_m)\right]^{\frac{3}{4}}(1 + L_i)A_i}{1 + \exp\left(\dfrac{-2V_m}{k'T}\right)\left[P_o(E_K)\right]^{\frac{3}{4}}(1 + L_i)A_i}\right]} \tag{7.12}$$

Values: $A_i = 0.75$ $\Delta G_n = 160 \, \text{meV}$

$$P_o = \left[1 + \dfrac{\beta_e}{\alpha_e}\right]^{-4} \qquad \beta_e = \exp\left(-\dfrac{V_m + 60}{26}\right) \tag{7.13}$$

α_e is equal to α_n in Table 7-1C with $\gamma_{\alpha n} = 1$ (no distortion).

$$L_i = \exp\left(\dfrac{-\Delta G_n}{kT} + \dfrac{E_K - V_m}{k'T}\right)\exp\left(-\dfrac{V_m + 60 + E_K}{9}\right) \tag{7.14}$$

$$D_o = \dfrac{1 + \exp\left(-\dfrac{V_m + E_K + 75}{k'T}\right)\left[1 - P_o(V_m + 35)\right]}{1 + \exp\left(-\dfrac{V_m - E_K}{k'T}\right)P_o(V_m + 35)} \tag{7.15}$$

Potassium channel current for Hodgkin-Huxley linear model:

$$I_K = \left[n(V_m + 21)\right]^4 (V_m - E_K)\left(\dfrac{1.17}{-E_K}\right) \tag{7.16}$$

n is the open probability using Hodgkin-Huxley rate constants (Table 6-1).

calculated, using the equation

$$n'(V_m) = \log\left(\frac{\Phi_o(V_m)}{\Phi_i(V_m)}\right) \log\left(\frac{M_o(V_m)}{M_i(V_m)}\right)^{-1}, \tag{7.17}$$

and the resulting curve for n', with $E_K = -75$ mV, is shown in Fig. 7-4.

Open channel probability

The potassium channel has a large γ-distortion factor, which reduces the electron-tunneling slope sensitivity from 1/26 to the 1/80 value of β_n. This distortion is caused by modulation of the open-gate energy barriers with the Ussing flux ratio parameters as described in Chapter 5. In addition to a reduced slope sensitivity, there is a 12 mV shift in the positive direction for both β_n and α_n rate constants. The open-channel probability (with the distortion removed) is given by Eq. 7.13.

7-4. The influx gating latch-up effect

Figure 7-3 clearly showed that gating of the potassium channel current causes a much greater attenuation (of the GHK curve) for the K⁺ influx than for the K⁺ efflux; this results in n' being greater than one. But why should the gating attenuation factor be different? The sodium channel has the same gating attenuation factor for both influx and efflux (Fig. 7-1). The reason for this difference is thought to come from the geometry of the channels and the location of control site q_1 (end of electron tunnel track).

A potassium influx ion has its average location, for crossing a gated energy barrier, shifted by a distance d' toward the outside of the membrane, as compared to the average location of an efflux ion. The ion exerts a strong force on the control site electron. The shift in angle makes it likely that a component of the influx ion's force on the control site electron acts to inhibit the electron from tunneling to site q_2. This keeps the energy barrier in the high state, blocking the ion's advance. In a simplified representation of this effect (Fig. 7-5), the inhibiting force on the control site electron is given as $F_e = F_1 \sin 2\theta'$. An estimate of the force and the associated electric field at the q_1 control site were calculated as follows:

$$F_e = \frac{\sin 2\theta'}{4\pi\varepsilon_0\varepsilon_r} \frac{e^2}{r_1^2} = 2.2 \times 10^{-11}\,\text{N} \tag{7.18}$$

$$E_{q1} = \frac{F_e}{e} = 1.4 \times 10^8\,\text{Vm}^{-1} \tag{7.19}$$

Estimated values: $r_1 = 8$ Å, $d = 1$ Å, $\theta' = 3.6°$, $\varepsilon_r = 2$

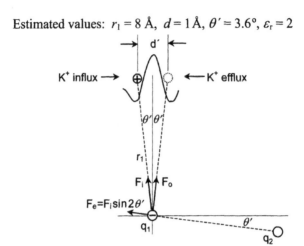

Fig. 7-5. Illustration of K^+ influx gating latch-up. The force F_i, from a K^+ influx ion, on the control site electron at q_1, has a component F_e acting along the direction of tunneling. This inhibits electron tunneling to site q_2, thus keeping the channel energy barrier in the high state and the gate closed. The force F_o from an efflux ion is at a right angle to the direction of electron tunneling and cannot inhibit tunneling to site q_2.

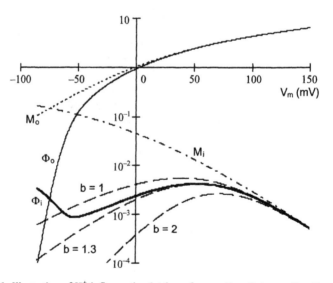

Fig. 7-6. Illustration of K^+ influx gating latch-up for one ($b = 1$) to two ($b = 2$) energy barriers. The electron-tunneling rate constant β_e is driven to high value. This substantially attenuates the influx by causing long dwell times for the electron at the control site and long gate-closed intervals.

The calculated electric field at the q_1 control site is equivalent to an 800 mV hyperpolarizing membrane potential! This is about -60 mV across the $q_1 - q_2$ tunneling sites, which places the rate constant β_e well into the saturated region of the curve (Fig. 6-2) for finite-range rate constants. With -60 mV across the tunneling sites, β_e equals about 100 ms^{-1} (using Eq. 6.1). The probability for the energy barrier to be in the open state is given by

$$P_{o1} = 1 - \frac{1}{1 + \dfrac{\alpha_e}{\beta_e}}, \qquad (7.20)$$

where

$$\beta_e \approx 100\,\text{ms}^{-1} \qquad \alpha_e = \exp\left(\frac{V_m + 60}{26}\right).$$

If only one of the energy barriers has gating latch-up, then it will be the one delaying the influx ion's transit through the channel and determining the influx attenuation. To observe the amount of attenuation for one or more energy barriers with gating latch-up, a graph (Fig. 7-6) was plotted with the influx calculated as

$$\Phi_i' = P_{o1}^b M_i, \qquad (7.21)$$

where b is the number of energy barriers with latch-up. As indicated by the curves, a single energy barrier seems to be involved in the influx attenuation. Two or more barriers would make the curve too steeply attenuated with voltage change. Only the β_e rate constant is subject to this latch-up and saturation. The first gated energy barrier encountered by the influx ion is the one, most likely, subject to latch-up; the repulsive forces from electrons in the adjacent tunnel tracks would have a smaller force component (if any) acting to repel the control site electron to site q_2.

The rate constant α_e, for the two-site model, gave the best results. This may be because the electric field from the nearby K⁺ influx ion is decreasing with $\sim r^{-2}$, causing the average charge for α_e to be skewed towards the q_2 site. This may be similar to the edge distortion (Section 5-6) of rate constant α_h for sodium inactivation.

Another way to describe the above is that three of the four modulated energy barriers obey the Ussing flux ratio test, but the first energy barrier encountered by the influx ion, determines that n' is greater than one. This energy barrier has the longest delay times for transiting influx ions and thus determines the overall magnitude of the influx and the magnitude of the flux ratio exponent. Thus, an exponent n' greater than one is due to a preferential attenuation of the influx by the gating latch-up effect.

The influx gating latch-up effect can cause an attenuation of the influx by more than an order of magnitude for negative membrane potentials, while producing a negligible effect on the efflux. Another characteristic of influx gating latch-up is that it produces a steep rectifying characteristic for the potassium channel. The steep rectifying characteristic of some inward-rectifying potassium channels is also likely produced by the gating latch-up effect; it would then be efflux-gating latch-up. Because the concentration of potassium ions in the cell is higher than the external concentration, efflux-gating latch-up could be a necessary mechanism to reduce outward flux and permit a net inward ion current in some inward-rectifying channels.

Chapter 8

FAR SITES, NEAR SITES AND BACK SITES

8-1. Ion channel mapping

Site maps were created (Fig. 8-1 and Fig. 8-2) to clarify how electron tunneling could account for gating and to illustrate the various tunneling distances associated with gating current and inactivation. These maps have a scale representation for the spacing between the sites, along the α-helix axis, and show approximate alignment with the cavities in the channel used for the modulated energy barriers. To simplify the site maps for illustration, the electron tunneling sites are shown along a straight line and the cavities for the energy barriers, which gate the ion current flow, are indicated by triangular displacements in the channel.

In addition to the main S4 tunneling site sequence, other arginine and lysine sites referred to as *far sites* are shown. They are part of the S4 protein segment that extends beyond the membrane into the cytoplasm. Surrounding the S4 are five other transmembrane segments. They act like a cage, protecting the far sites from contact with water. Contact with water molecules would stop the NH_3 inversion resonance and the amplification, due to energy loss to water molecules. The far tunneling sites of arginine and lysine account for the multiple time constants of the gating current and, combined with the gating cavities, they account for inactivation. A far site in the model usually has an associated inactivation-gating cavity.

In addition to far sites, some channels have a protein sequence with an arginine or lysine residue located towards the cell exterior, four or more amino acids from the q_1 activation control site. These tunneling sites are referred to as *back sites* because they are located back from the control site, near the outer edge of the protein. If they are not exposed to the aqueous exterior, or to water penetrating into their α-helix cavity, they would likely be amplifying electron-tunneling sites. Loss of amplification by exposure to water would substantially reduce the probability for electron tunneling to a back site. Back sites do not have gates, but may cause an increase in ion channel current with hyperpolarization by lowering the electron probability at the q_1 control sites. Increased ion channel current with hyperpolarization has been observed for some potassium channels (Ruppersberg et al., 1991) and in some cases it may involve the back sites. The electron-tunneling sites closest to the control site, in terms of tunneling time, are referred to as *near sites*. They are usually spaced every third residue and form the main S4 sequence.

The utility of site maps as a tool for understanding ion channel characteristics made it desirable to have a standardized format for assigning the tunneling site labels.

Format for assigning tunneling site labels

1. The arginine or lysine residue (on S4) nearest the activation gate is labeled q_1.

2. Moving toward the inactivation end of the protein, the q_n designation is incremented by one for each arginine or lysine, spaced every third or fourth residue until the end site for the sequence is reached.

3. Arginine or lysine sites located five or more residues from the sequence end site are labeled as far sites. The far site designation is q_{FX}, where X is the number of residues (residue #) from the S4 sequence end site. The difference between the residue # of two adjacent far sites equals the Rise #. The Rise # is cross-referenced to the tunneling distance and the time constant in Table 8-1.

4. Arginine or lysine back sites are generally located towards the outside edge of the protein S4 segment, five or more residues back from the q_1 control site. They are labeled as q_{BX}, where X is the number of residues from the control site.

5. The same sequence and direction, as in step 2, is followed for sites in S4 regions having only an inactivation gate (e.g. Na domain IV).

6. The same sequence and direction, as in steps 2 and 3, is followed for labeling sites in S4 regions having an activation gate and one or more inactivation gates, as in potassium channels.

7. The designation r_N is used to identify the location of residues with a negative partial charge responsible for open-gate barriers.

8. The cross sectional α-helix angle ϕ, listed in Table 8-1, is for the NH_3 tunneling site center. It is shown in Fig. 8-1 and Fig. 8-2 at the control sites, end sites, far sites and any back sites. It is aligned for zero degrees at the q_1 control sites. The angle helps in estimating the distance to a gating cavity in the channel. It is assumed that gating has been optimized by evolution, so that the α-helix orientation places the q_1 control sites at or near zero degrees, thus minimizing the distance and maximizing the ΔG value for the energy barrier.

Table 8-1. Estimated peak time constants for amplified electron tunneling between two arginine sites on an α-helix at 6°C

Rise #	$x_r(Å)$	$r(Å)$	$\theta(°)$	$\phi(°)$	$h_w(r)$	τ_{ep}	Note
0	–	–	–	0	–	–	–
1	1.5	6.27	76.1	100	23	0.7 ms	SB
2	3.0	8.38	69.0	–160	14	5 ms	SB
3	4.5	6.00	41.4	–60	25.2	0.5 ms	SJ
4	6.0	6.59	24.3	40	23	2 ms	SJ
5	7.5	10.58	44.8	140	9	60 ms	1C
6	9.0	11.33	37.4	–120	8	120 ms	1C
7	10.5	10.59	7.5	–20	9	60 ms	1C
8	12.0	13.04	23.0	80	6	1 s	1C
9	13.5	15.66	30.5	180	4	15 s	2C
10	15.0	15.84	18.8	–80	4	20 s	2C
11	16.5	16.56	4.8	20	3	1 min	2C
12	18.0	19.27	20.9	120	2	15 min	3C
13	19.5	20.88	20.9	–140	2	1.6 h	3C
14	21.0	21.18	7.4	–40	2	2 h	3C
15	22.5	22.85	10.0	60	1.8	16 h	3C
16	24.0	25.24	18.0	160	1.5	12 days	4C
17	25.5	26.22	13.4	–100	1.4	40 days	4C
18	27.0	27.00	0	0	1.4	100 days	4C

Note: SB: skirt backbone, SJ: space jump, 1C: one intervening crossing of α-helix
Sites marked SJ were calculated as a space jump. For calculations refer to the Appendix.

Sodium channel site map

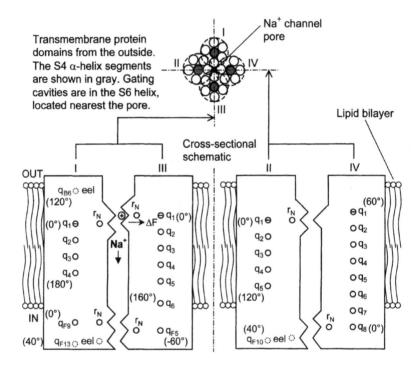

Fig. 8-1. The site map for sodium channel alpha subunit domains shows S4 arginine and lysine electron tunneling sites, designated by q_X, in relation to activation and inactivation energy barrier gates. An electron at a q_1 control site exerts a force ΔF on the sodium ion. This increases the energy required for the ion to pass the barrier, thus substantially lowering the transmission probability and resulting in a closed gate. Two additional energy barriers in sequence further reduce transmission probability and gate leakage when they are in the closed state. Far sites, designated q_{FX}, trap electrons (i.e. charge immobilization) and cause multiple long time constants for inactivation and gating current. The tunneling sites are for the giant axon of the squid *Loligo opalescens*. Additional sites for *Electrophorus electricus* (eel) are shown by dotted circles. The back site q_{B6} (for eel) with a time constant of tens to hundreds of milliseconds may alter the gating kinetics. To simplify mapping, the sites are shown along the S4 α-helix axis and have spacing proportional to the rise distance between the sites. The cross-sectional angle ϕ (Table 8-1) for the tunneling site on the α-helix is shown in parenthesis with 0° closest to the channel axis. The distance across the protein is ~45 Å, based on the total rise distance for the α-helix to cross the site map.

Potassium channel site map

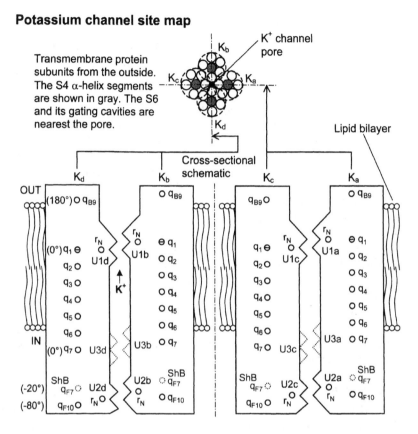

Fig. 8-2. The site map for potassium channel alpha subunits shows electron-tunneling sites in the S4 region. The activation gate at U1 is closed by the presence of an electron at one of the q_1 control sites. Polar residues (r_N) are oriented with their negative regions towards the gating site cavities. They provide the open-gate energy barriers for gating function and for ion selectivity. An inactivation gate is associated with the q_{F10} sites. The electron-tunneling distance of ~15.8 Å to these far sites gives a time constant in the range of seconds to minutes. This can account for the slow inactivation in delayed rectifier K^+ channels. The arginine and lysine sites shown are for *Loligo opalescens*. The *Shaker* B far sites at q_{F7} are also shown to illustrate their spacing. These sites have an inactivation time constant of tens to hundreds of milliseconds, resulting from a ~10.6 Å tunneling distance. The fast N-type inactivation (2–10 ms) of *Shaker* B results from closing the gate at U3 when the tunnel-track electron is at control site q_7. A cofactor residue L7 from the N-terminus region of the subunit is needed to close U3. The distance across the potassium channel protein is ~60 Å, based on the total α-helix rise distance across the site map. The back sites q_{B9} have their NH_3 tunneling sites located at 180° and thus are removed from any water penetration into the α-helix cavity. If these sites are amplifying, the activation gate at U1 would open the channel for a large hyperpolarization. The time constant could be about a second.

8-2. Far sites for inactivation, calcium signaling and memory

Protein sequences for many ion channels show arginine and/or lysine residues separated by four or more intervening residues from the cytoplasmic end of the main S4 arginine/lysine sequence (Figs. 9-1, 8-1 and 8-2). These are the far sites for electron tunneling. In the electron-gating model, the far sites play a major role in many aspects of ion channel function. They control inactivation gating when the time constants are longer than about 10 ms. They are responsible for the timing in the far-site calcium oscillator and thus control the calcium pulses that result in neurotransmitter release and muscle contraction. The far sites act as memory elements storing the electron and its charge for a period of time, which can range from tens of milliseconds to minutes, hours or possibly days. Far sites are usually near the cytoplasm and this allows an electron at a far site to be controlled by the local electric field from second messenger agents, such as calcium ions or by phosphorylation. Because of their importance, it was desirable to have a table showing estimated peak time constants and amplification for the various far-site tunneling distances. The values would be subject to several variables and depend on the characteristics of the intervening residues, but a nominal average value would allow a comparison of the electron tunneling times with experimental observations for gating current, inactivation times, calcium oscillator frequencies, etc.

Time constant and amplification values in Table 8-1

Table 8-1 shows the nominal peak time constant for an electron-tunneling distance r between two amplifying arginine tunneling sites. The peak time constants for the 6 Å and 6.59 Å space-jump distances can be calculated from Eq. 4.15 with f_{τ} equal to 1/2 and a temperature of 279°K. For other distances, the time constants were calculated using an equation that compensates for a $1/r$ decline and for the presence of the intervening amino acids (see Appendix). The peak-time constant also depends on the temperature and, to some degree, upon the amino acids between the tunneling sites. A distance-decay factor β is normally used in calculating electron transfer rate versus distance and for comparison with experimental data. The factor β was used instead of the theoretical terms $\beta_0\sqrt{U}$ in estimating time constants for electron tunneling to far sites. Measurements of electron transfer across proteins by Dutton and co-workers (Page et al., 1999; Moser et al., 1992) have shown that a β of 1.4 ± 0.2 Å^{-1} can represent a wide range of electron tunneling across proteins. Electron transfer across an α-helix has a predicted distance-decay constant of 1.26 Å^{-1}

(Winkler et al., 1999; Langen et al., 1995). The estimated time constant values in Table 8-1 are based on this distance-decay factor. The peak time constants follow a line having a rate (reciprocal time constant) versus distance slope, matching that of experimental data for tunneling across proteins. The amplification $h_w(r)$ declines with $1/r$ for a space jump and there would be an additional decline at longer tunneling distances because of the presence of non-amplifying intervening residues (see Appendix).

8-3. Near sites on the S4

The main part of the electron tunnel track consists of arginine and lysine amino acids, located every third residue on the S4. These are referred to as near sites, since they are near, in terms of tunneling time, to the activation control site. These are the principal sites determining the activation gating characteristics and the fast inactivation time constant. Rate curves generated from the Hodgkin-Huxley equations are accounted for, using only the near sites (Chapter 6).

Arginine versus lysine tunneling sites

The near-site spacing of every third residue gives the smallest possible tunneling distance for sites on an α-helix, and the smallest time constant (Table 8-1). It is interesting that most ion channels have arginine at the activation control site q_1 and at the adjacent site q_2. This suggests that arginine residues have some preferred property at this location. Lysine residues are more common as far sites, where time constants are longer. Lysine sites, with their higher inversion frequency, have less amplification ($h_w \sim 22$) than arginine and this would give a smaller time constant; however, a small difference in tunneling distance could override this and produce either a smaller or a larger time constant for the lysine sites. In this book the discussion of time constants is mostly about arginine residues because the amino acid sequences show arginine at the control sites and the adjacent sites, and the model is calibrated for this via the Hodgkin-Huxley equations. Lysine residues at these locations would likely produce a different time constant, perhaps differing by as much as a factor of five.

8-4. Back sites and hyperpolarization

In addition to far sites and near sites, protein sequences may have arginine or lysine amino acids that are five or more residues back from the activation control sites. A *back site*, in a voltage-dependent channel, is defined as an arginine or lysine residue located towards the outside edge of the S4 transmembrane protein segment and separated by four or more intervening residues from the activation control site. A back site that is also an amplifying electron-tunneling site, can act as a memory element, storing the electron and its charge for a significant period of time that depends upon the membrane voltage and the tunneling distance to the control site. Back sites can cause distortion to the ion current versus membrane voltage curve and may open the channel with hyperpolarization.

These sites may or may not be amplifying tunneling sites. If they are near the solution interface, they may be surrounded by water dipoles and the amplifying inversion resonance quenched. Figure 8-1 shows a back site for *Electrophorus electricus* sodium channel that could open one activation gate energy barrier on hyperpolarization and cause the channel to operate in an m^2 mode. If this is an amplifying electron tunneling site, there should be a measurable gating current for a large hyperpolarizing membrane voltage step.

Back sites in all subunits could cause a channel to open with a large hyperpolarization. The delayed rectifier potassium channel for *Loligo opalescens* has back sites, located nine residues from the q_1 control site. The NH_3 group is at 180 degrees, which would likely keep it protected from any water that might enter the α-helix cavity; however, it is close to the exterior interface and water there might make it inactive. There would be a force from any positive ions in the vestibule and gating cavity(s) acting to pull any electrons, at back sites, towards the control site, so that a large hyperpolarizing voltage would likely be needed to open the channel. Another reason that a large hyperpolarizing potential might be needed to open the channel is that the effective amplification is less. The amplification h_w would be reduced (Table 8-1) by intervening non-amplifying residues, but this would be partially compensated by the threefold increase (for a q_{B9} back site) in the distance x_r. The effective amplification would then be $3h_w(r)$, which could be about 12.

One reason a potassium channel might have amplifying back sites is that for a large hyperpolarizing potential they could disable the influx gating latch-up, thus enhancing the cation influx by more than an order of magnitude (Fig. 7-6). This would act to stabilize the resting potential and keep it from becoming excessively negative. In this fashion the potassium channel could act as a resting potential regulator for both polarities (around equilibrium potential E_K) by allowing a net inward potassium current.

Long lasting channel reopenings of a second or more, with hyper-polarization, might be accounted for by back sites. At a less negative potential, the far sites would be the likely cause of long lasting channel reopenings. Far sites and back sites can account for many of the reported ion channel characteristics having long recovery time constants.

8-5. Gating Current

Important clues about ion channel gating are provided by the magnitude and time response of the gating current to changes in membrane voltage. The ionic current depends on the movement of gating charges to open the gates and usually lags the gating current by at least a small time interval. Gating current has been extensively studied for sodium channels, starting with the pioneering work of Armstrong and Bezanilla (1974), and Keynes and Rojas (1974).

Recordings for gating current show three separate characteristics in response to a large depolarizing step in membrane potential:

1. Rapidly rising current phase

2. Rapidly decaying current phase having at least two time constants

3. Slowly decaying current phase having one or more time constants

In our model, the slow decaying current phase is caused by electrons tunneling to the far sites, where they become immobilized. The sodium channel of *Loligo opalescens* has a far site $q_{F5(III)}$ with τ_{ep} of 20 ms to 100 ms, and a far site $q_{F9(I)}$ with τ_{ep} of seconds to minutes. These sites can account for the slow decaying current phase.

The time constant of the rising current phase is likely due to a repulsive interaction between adjacent tunnel-track electrons. The first electron to leave the q_1 control site would tend to delay the others from leaving. As more of the adjacent electrons left q_1, the holding force on the remaining electron(s) would increase. Thus, the peak current could be delayed. This back holding force would likely increase the initial delay in the channel conductance versus time curve and could explain the initial delay for potassium activation (beyond n^4) observed by Hodgkin and Huxley (1952).

The rapidly decaying current phase can be analyzed with the SETCAP circuit model (Fig. 3-1). The model reveals a surprising characteristic. Instead of having discrete time constants, the time constant for the decaying gating current continuously increases. In the model, this occurs because a charge of almost $1e$ on the capacitor C_1 (representing site q_1) causes it to have a very low capacitance, making the initial time constant small. As charge is transferred to q_2 the capacitance and the time constant increase.

For the electron-gating model, equations were derived from the SETCAP circuit of Fig. 3-1*A* to determine the time response of the average gating current to a step change in membrane voltage. The gating current results from the transfer of electron average charge between two capacitors that represent an ensemble for two adjacent arginine tunneling sites. The determination of gating current, as a function of time, is treated as a standard circuit problem, except there is the additional complication that capacitance varies exponentially with voltage, and the total charge on the two capacitors is fixed at $1e$. Starting with the 2-site model of Fig. 3-1A, an equation was written for the sum of the voltage drops around the circuit loop.

$$\frac{i}{\text{þ}C_2(v_2)} + \frac{i}{\text{þ}C_1(v_1)} + Ri = v \tag{8.1}$$

Multiplying by the derivative operator þ and replacing current with $i = (v' - v)/R$, gives the differential equation

$$\frac{d(v'-v)}{dt} + \frac{1}{R}\left(\frac{1}{C_1(v_1)} + \frac{1}{C_2(v_2)}\right)(v'-v) = \frac{dv}{dt}, \tag{8.2}$$

where v' is the time varying potential across the two capacitors in series. Here we are interested in the response to a step change in the tunneling voltage v at time t_0. After time t_0, v has a constant value and $dv/dt = 0$. Because $C_1(v_1)$ and $C_2(v_2)$ are exponential functions of voltage, the equation was solved for time.

$$t(v') = -\int_v^{v'} \frac{RC_1(v_1)}{\left[1 + \frac{C_1(v_1)}{C_2(v_2)}\right](v'-v)}\, dv' \tag{8.3}$$

The function being integrated is equal to the time constant function (Eq. 3.6), but with a time varying voltage v' across the capacitors, resulting from a step change in the tunneling potential v. Rewriting the equation, incorporating the time constant function of Eq. 3.6 and letting $v = 0$ after time t_0, gives

$$t(v') = -\frac{RC_0}{\sigma}\int_v^{v'} \frac{\exp\left(-\dfrac{v'}{2H'}\right)}{\left[1+\exp\left(-\dfrac{v'}{H'}\right)\right]v'}\, dv'. \tag{8.4}$$

Let v be a voltage step, which goes from a negative resting value v at t_{0-} to $v = 0$ at t_0. The simplification of having the voltage step go to zero at time t_0, makes the potential v' equal to the potential across resistance R after time t_0. Curves were plotted for $t(v')$, using the following parameters for electron tunneling: $H' = 0.9$ mV, $RC_0/\sigma = 1$ ms at 6°C. The graph has

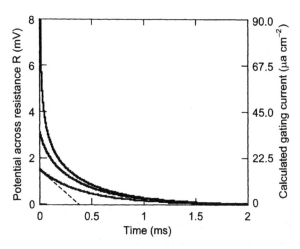

Fig. 8-3. The rapidly decaying current phase, resulting from a large depolarizing step in membrane voltage, has a continuously increasing time constant. This is illustrated by the curves for a time-dependent differential equation for the 2-site model (Eq. 8.4, Fig. 3-1A). The potential v' across resistance R was plotted for step changes in v from -8 mV, -3 mV and -1.5 mV to 0 mV. The potential across resistance R is proportional to the average current $i_d(t)$. The large step change from a resting potential of -8 mV up to 0 mV, causes a continuously increasing early time constant, from an initial value of about 40 μs up to 500 μs (6°C). The smallest step change of -1.5 mV to 0 mV causes an increasing time constant from about 370 μs up to 500 μs. Time constants were computed with Eq. 3.6. The initial value of ~40 μs was adjusted to reflect saturation as shown in Fig. 6-4B. The saturating value at 20°C was estimated to be 10 μs. The calculated gating current is calibrated according to Eq. 8.6. The -8 mV peak potential corresponds to a membrane holding potential 80 mV more negative than $V_{1/2}$. The gating current in the early portion of the curve (up to 200 μs) is about the same for the 4-site model, because the additional capacitors have negligible charge. After 200 μs, the time constant doesn't change much and the decay can be approximated as a single time constant.

the time axis as the abscissa (Fig. 8-3). It shows a continuously increasing time constant for the average gating current. The increasing time constant is caused by the exponential change in capacitance with voltage. If the capacitance remained constant, there would be a single time constant for the gating current decay. This is the case for a single tunnel track. The capacitance always has a constant minimum value when the electron is present at a tunneling site. When the electron is absent the capacitance has a maximum value. The curve for tunneling probability versus time, then follows an exponential curve (like the Hodgkin-Huxley time-constant curve) with a single voltage dependent time constant. It is the averaging of gating current for many ion channels that causes the continuously increasing gating-current time constant. The maximum variation in the time constant with time (Fig. 8-3) and with membrane voltage is limited to about $h_w/2$. This is because of the finite range of the β_e and α_e rate constants for electron tunneling (Fig. 6-2).

The gating current for a tunnel track with N sites is the sum of all the loop currents passing through the resistances R between each capacitor (Fig. 3-3).

$$i_g = \sum_{n=1}^{N-1} i_n \qquad (8.5)$$

For a large negative resting potential, most of the charge is on capacitor C_1 and when depolarization occurs, the average peak gating current is approximately equal to the peak current for i_1 (Fig. 3-3). A step in membrane voltage from a holding potential of $-80\,\text{mV}$ to $0\,\text{mV}$, for the sodium channel, corresponds to a peak voltage change of $v'_p = 8\,\text{mV}$ across the resistance R (with $\eta = 0.1$) and to an average peak gating current given by $i_{gp} = v'_p/R$. If $v'_p = 8$ mV and $R = 1.6 \times 10^{13}$ ohms at 6°C (Eq. 4.19), then the average peak gating current for a channel (with four electron tunnel-tracks) would be about 2×10^{-15} A. It was determined by Keynes and Rojas (1974) that the average value of the total gating charge displacement at saturation was, for the squid giant axon sodium channel, about 1882 charges per μm^2 (30 nC cm^{-2}). Dividing this by four charges per channel, gives a channel density of $d_c = 470$ channels per μm^2. Using these values, the peak gating current for an area of the membrane was calculated as

$$i_{gpa} = \frac{4v'_p \left(d_c \times 10^8\right)}{R} \approx 90\,\mu\text{A cm}^{-2}. \qquad (8.6)$$

The peak gating currents, calculated for the model, are in reasonable agreement with the observations of Armstrong and Bezanilla (1977) and Keynes and Rojas (1974), after taking into account differences in the voltage pulse amplitude and temperature.

8-6. Charge Immobilization

According to the electron gating model, when sodium channels of the squid *Loligo opalescens* are depolarized by gating pulses, tunnel-track electrons have a probability to enter the far sites q_{F9} in domain I and

q_{F5} in domain III (Fig. 8-1). Electron tunneling to these sites causes immobilization of the electrons and delays the full recovery of the gating current when the membrane is repolarized. There are two recovery time constants, corresponding to the $q_{F9} - q_4$ and $q_{F5} - q_6$ electron-tunneling distances. The recovery time from these far sites depends strongly on the repolarization voltage. It is evident from the model, that the maximum charge that can be immobilized is determined by the fraction of domains that have far sites. In domain IV, sodium channels have eight sites, spaced every third residue, but no far sites. If the domains I, II and III all have far sites, as does *Electrophorus electricus*, then the immobilized charge fraction is three-fourths. For the squid *Loligo opalescens*, the fraction is one-half (Fig. 8-1). A charge immobilization of about two-thirds was determined, from the gating current, by Armstrong and Bezanilla (1977) for the squid *Loligo pealei* sodium channel and inactivation was reported to recover along with the gating current. This suggests that there are far sites in domain I, II and III. If there had been far sites in only two domains (like *Loligo opalescens*), then the electron in the remaining tunnel track would have returned rapidly to the control site and closed the channel. The immobilized charge fraction, determined from the number of far sites, sets an upper limit for charge immobilization. Some types of ion channels may also have electrons immobilized at back sites, after a large hyperpolarization.

8-7. A calcium channel oscillator model using far sites

Voltage activated calcium channels control the passage of calcium ions into the cell where their concentration is normally very low. The low internal Ca^{2+} concentration facilitates Ca^{2+} signaling by increasing the dynamic range for oscillations in the local concentration. In this model, far sites establish the timing intervals for the low-frequency oscillations. Some characteristics that facilitate production of low-frequency oscillations are:

1. Low calcium concentration inside the cell
2. Appropriate far sites and inactivation gate locations
3. +2 charge for the calcium ion
4. Location of far sites near the cytoplasm

In the illustration of Fig. 8-5, the locations of arginine and lysine far sites, and the location of U2 inactivation cavities define a configuration for a calcium oscillator having a bursting mode. In this configuration, the electron in tunnel-track IV controls the fast pulsing, while the electron in track III controls the longer period bursting. The direction and time for electron tunneling is controlled by the electric field crossing the tunneling

Fig. 8-4.   L-Type Calcium Channels – S4 transmembrane segments

```
 qB7        q1            q5                    qF17   Domain I
ALGGKGAGFDVKALRAFRVLRPLRLVSGVPSLQVVLNSIIKA  Human cardiac muscle   (1)
ALGGKGAGFDVKALRAFRVLRPLRLVSGVPSLQVVLNSIIKA  Mouse cardiac muscle   (2)
ALGGKGAGFDVKALRAFRVLRPLRLVSGVPSLQVVLNSIFKA  Rabbit cardiac muscle  (3)
ALGGKGAGFDVKALRAFRVLRPLRLVSGVPSLQVVLNSIIKA  Rat cardiac muscle     (4)
ALGGKGAGLDVKALRAFRVLRPLRLVSGVPSLQVVLNSIFKA  Human intestinal muscle(5)
PMSSKGAGLDVKALRAFRVLRPLRLVSGVPSLQVVLNSIFKA  *Human skeletal muscle (6)
PMSSKGAGLDVKALRAFRVLRPLRLVSGVPSLQVVLNSIFKA  Rabbit skeletal muscle (7)
PAPGKSSGFNVKALRAFRVLRPLRLVSGVPSLQVVLNSIIKA  Bullfrog skeletal muscle(8)
     20°     0°         120°               20°

 qB11       q1            q5                    qF17   Domain II
KIMSPLGISVLRCVRLLRIFKITRYWNSLSNLVASLLNSVRS  Human
KIMSPLGISVLRCVRLLRIFKITRYWNSLSNLVASLLNSVRS  Mouse
KVMSPLGISVLRCIRLLRIFKITRYWNSLSNLVASLLNSVRS  Rabbit
KIMSPLGISCWRCVRLLRIFKITRYWNSLSNLVASLLNSLRS  Rat
KIMSPLGISVLRCVRLLRIFKITRYWNSLSNLVASLLNSVRS  Human
GAMTPLGISVLRCIRLLRLFKITKYWTSLSNLVASLLNSIRS  *Human
GAMTPLGISVLRCIRLLRLFKITKYWTSLSNLVASLLNSIRS  Rabbit
DIMSPLGISVLRCIRLLRIFKITRYWTSLNNLVASLLNSVRS  Bullfrog
    -20°     0°         120°               20°

            q1                        q8        qF11   Domain III
GSSAINVVKILRVLRVLRPLRAINRAKGLKHVVQCVFVAIRT  Human
GSSAINVVKILRVLRVLRPLRAINRAKGLKHVVQCVFVAIRT  Mouse
GSSAINVVKILRVLRVLRPLRAINRAKGLKHVVQCVFVAIRT  Rabbit
QSSAINVVKILRVLRVLRPLRINRAKGLKHVVQCVFVAIRT   Rat
QSSAINVVKILRVLRVLRPLRAINRAKGLKHVVQCVFVAIRT  Human
ESSAISVVKILRVLRVLRPLRAINRAKGLKHVVQCMFVAIST  *Human
ESSTISVVKILRVLRVLRPLRAINRAKGLKHVVQCVFVAIRT  Rabbit
ESSAISVVKILRVLRVLRPLRAINRAKGLKHVVQCLFVAIKT  Bullfrog
         0°                      -60°       -40

 qB7        q1            q5     qF5    qF13   Domain IV
EENSRISITFFRLFRVMRLVKLLSRGEGIRTLLWTFIKSFQA  Human
EENSRISITFFRLFRVMRLVKLLSRGEGIRTLLWTFIKSFQA  Mouse
EENSRISITFFRLFRVMRLVKLLSRGEGIRTLLWTFIKSFQA  Rabbit
EENSRISITFFRLFRVMRLVKLLSRGEGIRTLLWTFIKSFQA  Rat
EENSRISITFFRLFRVMRLVKLLSRGEGIRTLLWTFIKSFQA  Human
DESARISSAFFRLFRVMRLIKLLSRAEGVRTLLWTFIKSFQA  *Human
DESARISSAFFRLFRVMRLIKLLSRAEGVRTLLWTFIKSFQA  Rabbit
EESSRISITFFRLFRVLRLVKLLSRGEGVRTLLWTFIKSFQA  Bullfrog
     20°     0°        -140°     0°    80°
```

Amino acid sequences Accession (NCBI)

(1) Soldatov, 1992   Q13936   (5) Lyford et al., 2002   AAM70049

(2) Ma et al., 1992   Q01815   (6) Hogan et al., 1994   Q13698

(3) Mikami et al., 1989   P15381   (7) Tanabe et al., 1987   P07293

(4) Koch et al., 1990   P22002   (8) Zhou, et al., 1998   O57483

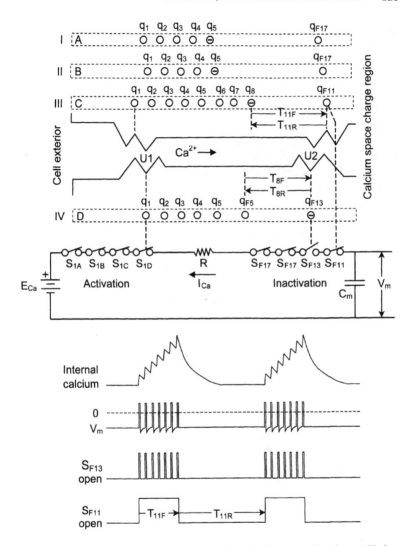

Fig. 8-5. The L-type calcium channels can produce low-frequency bursting oscillations of calcium. In the electron-gating model, bursting oscillations may be explained by electron tunneling to far sites and inactivation gating. When the channel opens (S_{F13} closes), calcium ions flood into the cell, depolarizing the membrane and causing a force on the tunnel track electrons. After an interval of T_{8F}, the track IV electron tunnels to q_{F13} closing the channel (S_{F13} opens). Nearby potassium channels then open and return the membrane to a negative potential. The negative membrane potential repels the electron back to the q_{F5} site. This again opens the channel and the cycle repeats. After multiple cycles, the accumulated depolarizing pulses (for time T_{11F}) cause the electron in track III to tunnel to q_{F11}, closing the channel (S_{F11} opens) and ending the burst mode for a time interval T_{11R}.

sites. This is determined by the membrane potential V_m, which alternates in polarity and also by the magnitude of local Ca^{2+} space charge. With a large hyperpolarizing membrane potential, the electrons would have a high probability to be at the q_1 control sites and the channel would remain closed. Above a threshold voltage, the channel would enter the negative conductance region and would then be triggered to the open state, whereupon calcium oscillation could follow.

Referring to the illustration (Fig. 8-5), when the channel opens (all switches close), calcium ions flood into the space charge region and V_m becomes depolarized. This causes the tunnel-track electrons to be attracted towards the space charge. After a time interval T_{5F}, the track IV electron tunnels to q_{F5}, thereby closing the channel (opening switch S_{F13}) and ending the current pulse I_{Ca}. The depolarized membrane triggers potassium channels (not shown) to open, which rapidly return the membrane potential to a negative value, causing a repulsive force on the electron. After a time interval T_{5R}, determined by the tunneling distance and the magnitude of the electric field, the electron tunnels to q_5 and again opens the channel. After multiple pulses in the bursting mode, lasting for the interval T_{11F}, the electron in track III, tunnels to q_{F11} and closes the channel for a time interval given by T_{11R}.

The peak time constants for L-type calcium-channel tunneling distances are listed in Table 8-2. The electron tunneling time would depend on the magnitude of the electric fields crossing the tunneling sites.

Table 8-2. Tunneling time constants for the calcium oscillator (Fig. 8-4)

Tunneling	Domain	r (Å)	$h_w(r)[x_r/4.5]$	Time	$\tau_{ep}(6°C)$	$\tau_{ep}(37°C)$
$q_{F5} - q_{F13}$	IV	13.04	16	T_8	1 s	40 - 130 ms
$q_8 - q_{F11}$	III	16.56	11	T_{11}	1 min	3 - 8 s
$q_5 - q_{F17}$	I, II	26.22	8	T_{17}	40 days	2 - 5 days
$q_5 - q_{F5}$	IV	10.58	15		60 ms	2 - 7 ms

The time constants for 37°C depend upon the temperature factor, which is likely less than the space-jump value of $Q_{10} = 3$. The τ_{ep} range shown is for Q_{10} ranging from 3 down to 2 (Eq. 4.13). The amplification decreases rapidly with increasing tunneling distance (r) as indicated in Table 8-1. This is compensated by the increase in potential (increase in η) between the two sites. For comparison with the $r = 6$ A reference, the effective amplification is expressed as $h_w(r)[x_r/4.5]$. It is thought that most L-type calcium channels would use the above format for oscillation; however, the frequencies might be quite different from one channel location to another.

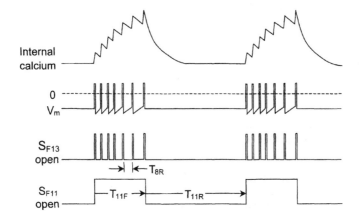

Fig. 8-6. Calcium channel bursting activity has been observed in some channels to have frequency adaptation. Here we consider the clearing of the space charge region. Each pulse injects calcium ions into the space charge region, which may only slowly be cleared. The calcium ions causes a local attractive force on the tunnel-track IV electron, which increases the time T_{8R} for the electron to tunnel back to the site q_{F5} and open the channel. The frequency adaptation shown here is caused by the rise in the average local calcium concentration, which delays the return of the tunnel track IV electron to q_{F5}. The channel cannot open, until the electron leaves the q_{F13} site.

Externally triggered calcium oscillations - ELF magnetic fields

When the resting membrane potential is below the oscillation threshold, L-type calcium channels can act like a sensitive regenerative receiver and be triggered to pulse by incoming calcium oscillations having a matching frequency, or they can be triggered to pulse by an incoming action potential.

An external source that has frequently been observed to trigger or alter calcium oscillations is that of low frequency magnetic fields. A particularly sensitive frequency for triggering oscillations by magnetic fields is at about 10 Hz. This approximately matches the frequency for electron tunneling between the sites q_{F5} and q_{F13}. Tunneling between these sites is therefore a likely location for inducing calcium oscillations at this frequency. A number of publications have described unusually high cellular sensitivity to low frequency magnetic fields with detection levels well below the thermal noise level. These ELF magnetic fields have been reported to couple to calcium oscillations and they are apparently amplified (Galvanovskis and Sandblom, 1997). In the far-site calcium oscillator model, regenerative feedback could provide some additional amplification. A calcium oscillator biased at the threshold of oscillation would have an enhanced sensitivity to ELF magnetic fields.

A sleep mode (i.e. long-term inactivation) for L-type calcium channels

Another interesting characteristic of L-type calcium channels relates to the q_{F17} far sites in domain I and II. The peak time constant for electron tunneling across this distance is listed as 2-5 days at 37°C in Table 8-2; however, the electric field crossing the sites could reduce the tunneling time. These sites most likely have associated gating cavities, which would cause inactivation when an electron tunneled to the q_{F17} site. One observation suggesting inactivation is that all residues in domain I and most residues in domain II, between and including q_5 and q_{F17}, are identical in the listed species (Fig. 8-4). This suggests that the region has an important function, such as inactivation. The homology would keep the electron tunneling time to q_{F17} about the same for all the listed species.

These inactivation gates would cause the calcium channels to shut down for an extended period when electrons tunneled to the q_{F17} sites. Tunneling might occur after a day or so of active pulsing. It seem that nature has built-in a resting period for these channels and their target, the muscles, and requires a rest period (perhaps 8 hours) with a hyperpolarized membrane potential for recovery. From the model, it is inferred that a sleep time is required to reset the electrons in calcium-channel far-sites back to their normal functioning state. It is thought that these long time-constant inactivation gates could be associated with the sleep cycle.

A far site mutation that could alter calcium oscillations

In reviewing the amino acid sequences for the calcium oscillator model, one L-type calcium channel for human skeletal muscle caught my attention. This is shown in Fig. 8-4 with an asterisk. In this channel, the q_{F11} site in domain III is serine, instead of the usual arginine. This is a R926S mutation. It seemed that this sequence might be from a patient with symptoms like hypokalemic periodic paralysis, as indicated by several references to this condition; however, other mutations known to cause hypoPP are not present in the alpha-1s subunit sequence. Other mutations known to cause hypoPP are for R528H and R1239H, occurring in domain II and IV (Ashcroft, 2000). The R528H mutation replaces arginine at the $q_{1(II)}$ control site with a histidine residue, which is reported to shift the $V_{1/2}$ voltage. The R1239H mutation replaces the $q_{2(IV)}$ residue and also shifts the $V_{1/2}$ voltage. In contrast, the mutation R926S would eliminate the bursting mode for the calcium oscillator by eliminating electron tunneling to the q_{F11} site. This would likely cause continuous pulsing and increase calcium entry into the cell. Based on the calcium oscillator model, one could think that this mutation might cause adverse medical symptoms.

Chapter 9

ELECTRON-GATED K⁺ CHANNELS

Many experimental observations have been reported on the properties of *Shaker* potassium channels. This, combined with the simplification that the potassium ion channel has four identical subunits, makes this an interesting channel on which to focus attention. There are, however, many subtleties in the experimental observations for this channel, and one objective here is to seek an explanation for some of these observations, based on the electron-gating model.

One of the benefits of the electron-gating model is that it introduces a vehicle to explain a wide range of observed electrostatic effects and the results of many experiments. Electrostatic interaction involving ions, pore blockers, toxins, and charges inside and outside the channel can alter the electron probability at the activation and inactivation control sites; can alter time constants and can cause charge immobilization in the far sites, or possibly the back sites. There is also an unusual characteristic for potassium channels, which causes substantial attenuation of the influx and some surprising characteristics for external toxins and pore blockers. This is the *influx gating latch-up effect*, described in Section 7-4. With all of these characteristics, a powerful model is needed to account for the numerous seemingly unrelated observations.

It seems appropriate to begin this chapter by describing the electron-gating model for the outward-rectifying potassium channels, in particular

Shaker and *Loligo opalescens* – the potassium channel characterized by the rate constants of Hodgkin and Huxley. The potassium channel geometry presented is considered to be the most likely configuration, given the constraints imposed by the electron-gating model, the amino acid sequence, and published experimental observations.

9-1. Activation and inactivation of K_v channels

A potassium ion channel site map for *Loligo opalescens* is shown in Fig. 8-2. It is based on the amino acid sequence determined by Rosenthal and Gilly, 1996. There is a lysine far site at q_{F10}, which has a tunneling distance to site q_7 of ~16 Å (Table 8-1), corresponding to a time constant of seconds-to-minutes. This far site can account for the slow inactivation first described by Ehrenstein and Gilbert (1966).

The amino acid sequence for *Shaker* B (Pongs et al., 1988) is the same, in the far-site region of S4 as *Loligo opalescens*, except for the glutamine residue (Q) at site q_{F7} (Fig. 9-1 and Fig. 9-2). For *Shaker* B, the arginine (R) residue at site q_{F7} disables the $q_{F10} - q_7$ electron-tunneling time constant and replaces it with a $q_{F7} - q_7$ time constant of 50 ms to a second, corresponding to the ~10.6 Å tunneling distance. There is an even shorter time constant of 2 to 10 ms for q_7, acting as an inactivation control site. This is similar to the sodium channel inactivation time constant with its inactivation control site q_8. The inactivation control sites need adjacent displacement structures in the channel to have modulated energy barriers for inactivation gating. The displacement structure associated with site q_7 apparently requires the L7 residue from the N-terminus region of the subunit and the A463 residue of S6 to facilitate inactivation. According to mutation experiments (Hoshi et al., 1990 and 1991), neutralizing either the L7 residue from the N-terminus region of the subunit, or the A463 residue of S6, removes the fast N-type inactivation. Figure 9-1 shows an alignment, which correlates with findings of these mutation experiments.

It has also been reported that fast N-type inactivation is not voltage dependent (Hoshi et al., 1991). Increasing the membrane depolarization increases the probability for electron tunneling from site q_7 to the far site q_{F7} and from q_{F7} to q_{F10}. If these three sites have gates, then a change in membrane voltage will change the gate-open probability for each of these sites. For the resulting product of the probabilities, the changes may cancel. An increase in membrane depolarization increases the electron and gate-closed probability at q_{F10}, but this is offset by a decrease in the

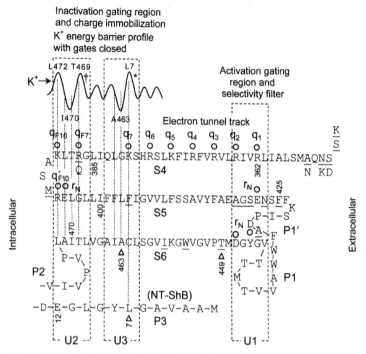

* q₇   Fast inactivation (N-type) – needs the N-terminus residue L7 and A463
† q_F7   Medium inactivation (C-type) – residue 387 must be R or K
† q_F10   Slow inactivation (C-type) – residue 387 must not be R or K (i.e. no q_F7 site)

Fig. 9-1. Amino acid alignment diagram for K⁺ activation and inactivation gating. The sequence of *Shaker* B potassium α-subunit is shown for regions including S4 through S6. Residue differences with *Loligo opalescens* are underlined and (some) *Loligo* amino acids are indicated below. The N-terminus (NT-ShB) is a cofactor needed for expressing fast, N-type inactivation. It is inserted as a liner in the channel as shown above. The residue L7 is aligned with tunneling site q₇ to account for a reported L7 hot spot. Neutralizing L7 removes the fast (2–10 ms) N-type inactivation (Hoshi et al., 1990; Zagotta et al., 1990). Alignment of S6 was adjusted so that the reported hot spot at A463 would also be in alignment with q₇. Neutralizing A463 also removes the N-type inactivation (Hoshi et al., 1991). The q₇ (N-type) inactivation is similar to q₈ inactivation for the sodium channel. With N-type inactivation disabled, the C-type inactivation, with a time constant determined by the q₇ to q_F7 tunneling distance, becomes dominant. It has a time constant of 50ms to 1s. For *Loligo opalescens*, the arginine site at q_F7 is replaced by glutamine (Q). This increases the tunneling distance from 7 to 10 residues and results in a time constant of seconds to minutes and a slow C-type inactivation. The sequence LAITL forms the gating cavity (U2) for C-type inactivation. The gate can be closed by the electron at either control site q_F7 or q_F10. The above gating configuration causes the selectivity filter residues (VGYG), at the end of S6, to be in approximate alignment with the q₁ control site. This alignment is essential for activation gating.

Fig. 9-2. Amino acid alignment for K⁺ gating

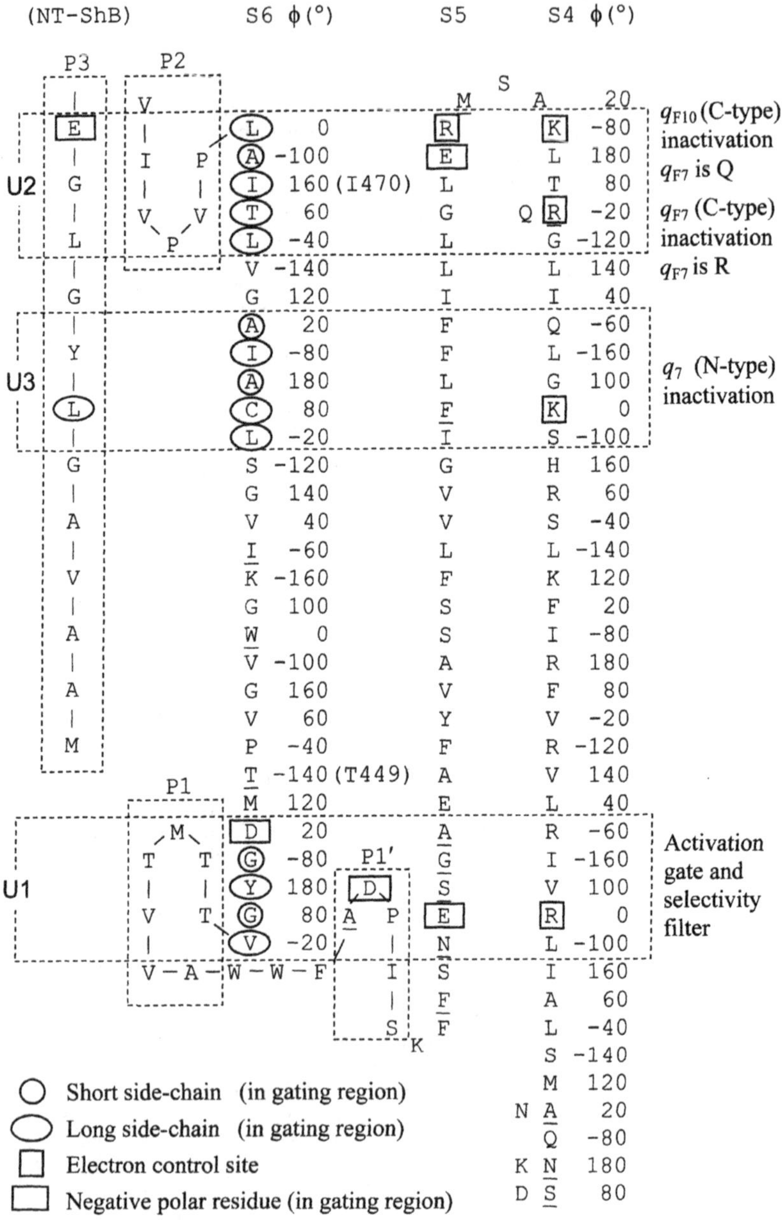

(NT-ShB)	S6 $\phi(°)$	S5	S4 $\phi(°)$	

P3   P2

				S			
	V			M	A	20	q_{F10} (C-type)
E		L	0		R	K −80	inactivation
I	P	A	−100		E	L 180	q_{F7} is Q
U2 G		I	160 (I470)		L	T 80	
L	V V	T	60		G	Q R −20	q_{F7} (C-type)
L	P	L	−40		L	G −120	inactivation
		V	−140		L	L 140	q_{F7} is R
G		G	120		I	I 40	
		A	20		F	Q −60	
Y		I	−80		F	L −160	
U3		A	180		L	G 100	q_7 (N-type)
L		C	80		F	K 0	inactivation
		L	−20		I	S −100	
G		S	−120		G	H 160	
		G	140		V	R 60	
A		V	40		V	S −40	
		I	−60		L	L −140	
V		K	−160		F	K 120	
		G	100		S	F 20	
A		W	0		S	I −80	
		V	−100		A	R 180	
A		G	160		V	F 80	
		V	60		Y	V −20	
M		P	−40		F	R −120	
	P1	T	−140 (T449)		A	V 140	
	M	M	120		E	L 40	
	T T	D	20	P1'	A	R −60	Activation
U1		G	−80		G	I −160	gate and
		Y	180	D	S	V 100	selectivity
	V T	G	80	A P	E	R 0	filter
		V	−20		N	L −100	
	V − A − W − W − F			I	S	I 160	
					F	A 60	
				S	F	L −40	
				K		S −140	
						M 120	
					N A	20	
					Q	−80	
					K N	180	
					D S	80	

○ Short side-chain   (in gating region)
◯ Long side-chain   (in gating region)
□ Electron control site
▭ Negative polar residue (in gating region)

Shaker B and *Loligo opalescens* (underlined where different) electron and gate-closed probability at q_{F7} and q_7. It is these opposing voltage sensitivity factors, at the three electron-tunneling sites, that apparently cause the fast N-type inactivation time constant to be voltage independent. An arginine residue (site q'_{F10}) on S5 may also be an amplifying electron-tunneling site. Electron tunneling between q'_{F10} and q_{F10} would not be very sensitive to membrane voltage changes, since the tunneling direction is approximately at right angles to the electric field. With prolonged depolarizations, all of the tunnel-track electrons can become immobilized at the far sites of the four identical potassium subunits. This differs from sodium channels where, at most, two or three of the four tunnel-track electrons can become immobilized at far sites.

Calibration for angles listed in Fig. 9-2

The angles shown for the S4 α-helix residues in Fig. 9-2 were calibrated to have $0°$ at the q_1 control site. This $0°$ reference is for the tunneling site at the end of the arginine or lysine side chain to be at the closest point to the central axis of the channel. The angles for S6 residues are for the residue centers on the backbone. These angles were calibrated to have the middle residue of cavities U1 and U3 to be at $180°$, the furthest point from the channel axis. The location of T449 at $-140°$ is for the residue center on the backbone. For the mutation T449K with a long lysine side chain, the NH_3 group at the end could project towards the channel and be exposed to water in the pore (or to water in the DMTPV cavity). This could give it some probability to have an excess positive charge.

9-2. Structural constraints for activation gating

As shown in Fig. 9-1, there are about 15 residues linking the S5 α-helix to the S6 α-helix. This short chain of residues will be referred to as H5. The requirements for the activation gate of the electron-gating model are as follows:

1. The gating cavity U1, consisting of sequence VGYGD, must be in (approximate) alignment with the (R362) q_1 control site on S4.
2. Residues forming the pore loop (P1) should block ions and water from passing through the center of the pore. This is needed to maintain the large potential drop across the selectivity filter as ions and water travel in a region defined by the activation-gating cavity in each of the four subunits. It has been reported that 80% of the transmembrane electric potential falls across the eight amino acids between residues 441 and 449 (Yellen et al., 1991).

3. Residue D431 should be located near the bottom of the gating cavity U1 so it can contribute to the electric field (from r_N) and the open-gate energy barrier.

4. To have agreement with TEA$^+$ experiments, residues T439 and M440 should be near the inside edge of the P1 pore loop.

5. To have agreement with experiments, residue W434 should be in a position to block ion current for a W434F mutation.

These constraints resulted in the pore loop configuration shown in Fig. 9-1 and Fig. 9-2. In the figures, the H5 linker between the S5 α-helix and the S6 α-helix is split into two parts: P1 for the pore loop and P1′ for a back loop, which places residue D431 near the bottom of the gating cavity. The link between P1 and P1′ consists of residues F433, W434 and W435. This link might pass in front of S6 or wrap around near the end of it.

When the four potassium subunits join together to form the gating region and selective filter, they are not in exact alignment along the pore axis. To have an activation gate with a characteristic of n^4, the potassium ions must pass sequentially through four gating cavities. There is one U1 gating cavity for each subunit, and the ion visits each cavity in sequence by traveling counter-clockwise (opposite the α-helix advance) as it advances along the channel. Each subunit is incremented by a distance d_z, which is shown in Fig. 9-4 as one rise distance (1.5 A); but d_z could have a different value, depending upon the bonding between the subunits. The above arrangement leaves a key residue (T449) just beyond the U1 gating cavity, towards the channel interior.

In mutation experiments, it has been shown that the sensitivity to external TEA$^+$ and α-KTx was virtually abolished by making residue T449 positive (T449K). Conversely, the sensitivity to TEA$^+$ was enhanced 50-fold by making it aromatic, T449Y or T449F (MacKinnon and Yellen, 1990; Heginbotham and MacKinnon, 1992). The above observations for sensitivity to external TEA$^+$ and scorpion toxins, along with site mutations at T449, have suggested that residue 449 defines the external edge of the pore, with the selectivity filter sequence GYGV folding back towards the cytoplasm. The electron-gating model cannot function with this configuration. This configuration does not allow alignment of the q_1 control site with the gating cavity, and there is no apparent way to have activation gating from the q_1 control site. This leaves us with the question: how can the observations for sensitivity to external TEA$^+$ be explained with residue T449 located inward from the selectivity filter as in Fig. 9-1 and Fig. 9-2, and as suggested by recent studies (Andalib et al., 2004)?

9-3. Influx gating latch-up and TEA⁺ sensitivity

According to the influx gating latch-up effect, developed in Section 7-4, external TEA^+ would have a strong interaction with the tunnel-track electron for the outermost control site q_{1a}. This interaction would cause latch-up, increasing the probability for the electron at q_{1a} to near unity, which in turn would hold TEA^+ in a blocking position at the entrance to the selectivity filter. Only the outermost control-site electron experiences the latch-up because of the angle of the applied force (Fig. 7-5 and Fig. 9-6B) and the lack of other counteracting forces. Only a small amount of external TEA^+ is required to block the channel, because of the increased probability of the electron at q_{1a} and its holding force on a positive charge at the entrance to the selectivity filter.

If T449 is located as shown in Fig. 9-1 and is mutated to a positively charged residue (T449K or T449R), then a counteracting force would be applied to the control site electron, which could substantially reduce the electron probability at q_{1a} and remove the latch-up. The magnitude of the holding force on a positive charge at the entrance to the activation-gating region and selectivity filter depends on the distance x_a, which determines the angle of the force (Fig. 9-6B). *In the model, it is the influx gating latch-up effect, not residue T449, that binds external TEA⁺ to the mouth of the pore.* The presence of external TEA^+ has been reported to slow C-type inactivation. This is because, with TEA^+ bound to the mouth of the pore, there is a strong attractive force exerted on the tunnel-track electrons, which reduces their probability and dwell time at the inactivation control sites. This reduces the probability for closing the inactivation gates.

Another observed effect for a mutation T449K or T449R was to reduce single channel currents. This would likely be due to electrostatic repulsion by the positive charge on residue 449. Mutations that produce no change in charge might also cause some change in TEA^+ sensitivity by an indirect electrostatic effect, such as altering the probable location of a nearby K^+ ion, based on differences in the geometry of the 449 side chain. One might think of residue 449 as being part of another cavity in the S6, formed by residues DMTPV. It would not be a gating cavity, since there is no control site for the electron aligned with it and the maximum steady-state probability for the electron at q_2 is less than 0.2, since it is not an end site. However, there could be a sufficient field to direct ions and water dipoles into the cavity, and that could make the charge distribution sensitive to the side-chain geometry (and perhaps to the hydrophobicity) of residue 449. With four electrons having a significant probability to be at their q_1 control sites, one could expect that several positive charges

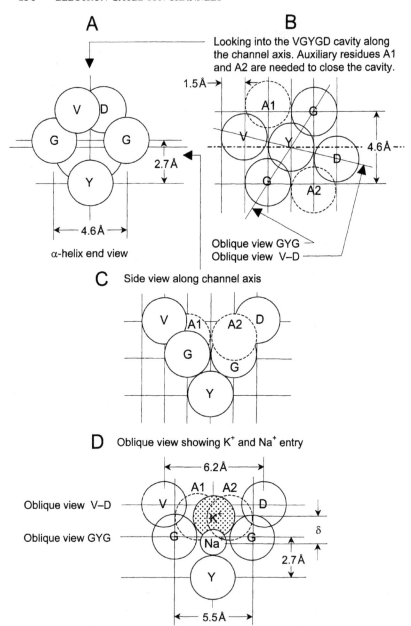

Fig. 9-3. Four views illustrate the K^+ selectivity filter and the activation-gating cavity. Dimensions are based on a 2.3 Å backbone radius and a 1.5 Å rise per residue distance. The threonine (T) residue from the sequence TVGYGD is likely located near A1, which would help close the cavity, as would the outward projecting side-chains (not shown). Distance δ (in Fig. D) determines the Na/K selectivity ratio for the cavity.

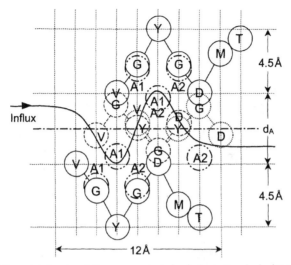

Fig. 9-4. The potassium ion follows a punctuated spiral counter-clockwise path, visiting the U1 gating cavity (VGYGD) in each of the four potassium subunits. In coming together to form a narrow passageway, each of the potassium subunits advances by a distance d_z (~1.5 Å), which corresponds to one-quarter turn about the channel axis. The dashed circles represent residues of the near subunit; the dotted circles are on the far subunit. The pore loop residues (not shown) block ion passage along the channel axis.

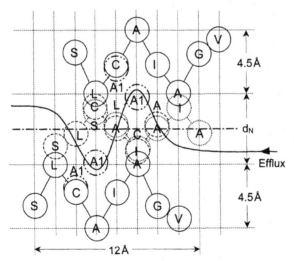

Fig. 9-5. The gating region U3, for N-type inactivation, has the same geometry as the activation gate at U1, but with different amino acids. The residue L7 from the N-terminus of each subunit has been determined to be a cofactor for fast inactivation (Hoshi et al., 1990). Residue L7 likely acts to close the U3 gating cavity at location A1. This would increase the energy barrier by requiring greater ion displacement against the electric field from an electron at the inactivation control site (q_7).

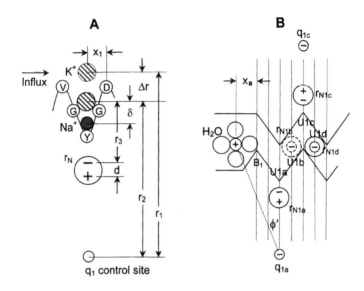

Fig. 9-6. As a potassium ion moves through the gating region and selectivity filter, it visits sequentially four energy wells. Four modulated energy wells are required to have the n^4 characteristic of the Hodgkin and Huxley model. An energy well is caused by a displacement of the ion into a cavity (VGYGD) off the channel axis, towards the corresponding q_1 control site. Dipoles (r_N) below each cavity, represent the equivalent electric field from nearby aspartate (D) and glutamate (E) residues. This short-range field determines the magnitude of the open-gate energy barrier. Before an influx ion can enter the first energy well, water molecules must be displaced from the ion. A strong force from the four control-site electrons plus a weaker force from the r_N dipoles, acts on the ion pulling it into the first gating cavity. Rough calculations indicate that this force should be sufficient to strip away the several water dipoles that keep the K^+ ion from entering the cavity. The removal of the water dipoles from the ion becomes part of the overall energy balance and force balance equations between the four electrons and three or four ions that have a high probability for being simultaneously in the channel. The energetics differ from current theory because of the presence of the four gating electrons inside the membrane protein, but the requirement for an overall energy balance is the same.

would be in the channel, so as to have an overall force and energy balance.

Another aspect of the influx gating latch-up effect is that the holding force is sensitive to the radius of the ion or toxin. Ions or toxins having the same charge but with a larger radius, would have a larger distance x_a (Fig. 9-6B) and a larger holding force from the electron at q_{1a}. TEA$^+$ is too large a molecule to pass through the selective filter, but ions such as rubidium can. The rubidium ion has a slightly larger radius than the potassium ion, but the influx current is substantially less because the

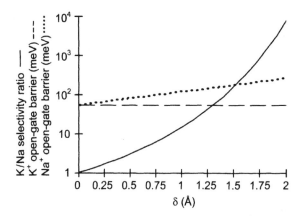

Fig. 9-7. The K/Na selectivity ratio increases rapidly with Na^+ penetration (δ) into the gating cavity. Calculations for selectivity (Eq. 9.2) were normalized for an average open-gate energy barrier of 54 meV for K^+. A selectivity ratio of 1000 was obtained for a sodium ion penetration (δ) of 1.5 to 2 Å beyond the potassium ion in the gating cavity.

larger x_a and holding force increases the probability for the electron to be at q_{1a}, which gives the energy barrier a higher probability to be in the high-energy closed state. The ion can only pass the barrier during the brief intervals when the electron is not at the q_{1a} site, but with hyperpolarization there is some probability for gate leakage in the closed state.

9-4. K/Na selectivity ratio

Current theory for K^+ selectivity, in discriminating against the smaller sodium ion, is based on the difference in the radius of the two ions; however, the theory requires holding a tight dimensional tolerance on the distance between subunits and it is not compatible with incremented potassium subunits as required by the electron gating model.

With the introduction of gating cavities, a different mechanism for ion selectivity becomes possible. A substantial discrimination against the smaller sodium ion can be achieved if the sodium ion penetrates into the gating cavity a distance δ beyond the potassium ion penetration (Fig. 9-7). Calculations were made (Eq. 9.2) to determine the approximate magnitude for a K/Na selectivity ratio as the distance δ was increased from zero (Fig. 9-7). As δ increases, the K/Na curve becomes steeper, reflecting the rapid change in electric field intensity from the equivalent dipole r_N (Fig. 9-6). These calculations are only approximate, but they indicate that a K/Na selectivity ratio of 1000 could be achieved in a single gating cavity for a displacement δ of 1.5 to 2 Å. The selectivity ratio defined by Eq. 9.2 is the ratio of the two rates at which ions cross an open-gate energy barrier in a single gating cavity. Since this method of selectivity is determined by the cavity within each subunit, it is compatible with having incremented subunits and multiple gating cavities.

Selectivity calculations for a gating cavity

Energy barriers:

K^+ open-gate barrier:

$$G_{oK} = 54 \text{ meV} \qquad \text{(from match-up in Fig. 5-6)}$$

Na^+ open-gate barrier:

$$G_{oNa}(\delta) = \frac{zee_0 10^{13}}{4\pi\varepsilon_0\varepsilon_r}\left(\int_{r_3-\delta}^{r_3+\Delta r} \frac{1}{r^2}dr - \int_{r_3-\delta+d}^{r_3+\Delta r+d} \frac{1}{r^2}dr\right) \qquad (9.1)$$

(This is similar to Eq. 5.10, but for a dipole; energy is in meV.)

Values used for calculation:

$d = 2$ Å, $\Delta r = 2$ Å, $r_3 = 4$ Å, $e_0 = 1$, $\varepsilon_r = 2$, $z = 0.18$
$\varepsilon_0 = 8.85 \times 10^{-12}$ $C^2N^{-1}m^{-2}$, $e = 1.602 \times 10^{-19}$ C, $kT = 24$ meV
(Fig. 9-6A is the reference, z is the dipole partial charge fraction)

K/Na selectivity ratio:

$$S_R(\delta) = \frac{\exp\left(-\dfrac{G_{oK}}{kT}\right)}{\exp\left(-\dfrac{G_{oNa}(\delta)}{kT}\right)} \qquad (9.2)$$

A selectivity ratio curve was plotted in Fig. 9-7 using the above values. A similar curve was obtained using the following alternate values:

$$d = 1 \text{ Å}, \quad \Delta r = 3 \text{ Å}, \quad r_3 = 3.55 \text{ Å}.$$

The displacement Δr was originally estimated to be 2 Å with Eq. 5.11, but the geometries of Fig. 9-4 and Fig. 9-5 suggest the value might be as much as 3 to 4 Å.

Selectivity for multiple cavities

Each of the subunits would have about the same selectivity ratio, but how should the overall selectivity be computed? The analysis for flux gating (Section 7.3), indicates that the overall selectivity ratio would depend on whether the channel is in the fully open state or in a region of significant attenuation by gating. Without gating, the channel current fluxes can be described, approximately, by the GHK relation over the full range of membrane voltage. When the channel is gated, the GHK relation is followed only when the membrane voltage is near that for the fully open channel. In the other region there is attenuation of the influx and efflux as a result of gating (Fig. 7-3).

The time required for ions to transit across the channel energy barriers (which determines the flux) depends strongly on gating. For a fully open channel, the time delay for an ion to cross multiple barriers is additive and, with a forward driving force on the ions, transiting the four activation gating cavities (U1$_a$ through U1$_d$) would give a 4x increase in the transit time for both sodium and potassium ions, thus resulting in the same overall selectivity ratio as for a single cavity.

When the channel is gated and the open-channel probability is small, there is a narrow window in time, when all of the gated barriers are in the open state. During this narrow window, the probability for an ion to cross all of the barriers (with a forward driving force on the ion) is the product of the probabilities to cross each barrier. The probability of crossing the energy barrier in the fixed time window is proportional to $\exp(-G_0/kT)$ and with 4 barriers it would be proportional to $\exp(-4G_0/kT)$. This would substantially increase the overall Na/K selectivity ratio.

In addition to U1, the inactivation-gating cavity U2 has a nearby negative dipole, in each subunit, which would likely produce an open-gate energy barrier and some selectivity against sodium efflux ions. The inactivation gating cavity U3, however, has no nearby negative dipoles and would not have a significant open-gate energy barrier. With an open channel, the ions would not be pulled into the U3 cavities. If the N-terminus chains, for the *Shaker* channel, were absent from the interior, many ions might then bypass the U3 cavities and transit through the center region of the channel.

9-5. C-type inactivation gating

For *Shaker* B, there are two inactivation control sites (q_{F7} and q_{F10}) in the far-site region of S4, both using cavity U2 for inactivation. When the tunnel-track electron is at either of these sites, the field from the electron causes an increase in the energy barrier for an ion at the U2 gating cavity. The inactivation time constant is determined by the tunneling distance q_7 to q_{F7}, but this time constant, while being present in the gating current, is not apparent for the ion channel current, because of the much smaller time constant for N-type inactivation. The electron arriving at q_7 closes the channel first, so when it tunnels to q_{F7} and then rapidly to q_{F10}, the channel remains closed. Neutralizing or removing the N-terminus of *Shaker*, reduces or eliminates the gating energy barrier and this increases the inactivation time constant to a value determined by the q_7 to q_{F7} tunneling distance.

The gating cavity U2 shown in Figs. 9-1 and 9-2, for C-type inactivation, is similar to the U1 and U3 gating cavities, but with different residues. Based on the residue angles, there is only one location at which the U2 cavity can be located. This is defined by the residues LAITL, which gives the cavity a tilt of 20°. The protein sequence for *Shaker* B (Pongs et al., 1988) shows arginine (q_{F10}) and lysine (q_{F7}) residues in a position to use a cavity at this location for gating. In addition, a number of studies have suggested that this is where a C-type inactivation gate is located. Mutations at T469 significantly enhance or decrease the affinity for internal TEA$^+$ (Choi et al., 1993). Mutations at I470 have suggested that this residue is in the center of a cavity and is the binding site for internal TEA$^+$ (Melishchuk and Armstrong, 2001).

As with the U1 and U3 cavities, an additional residue is likely needed to close one side of the cavity, and to permit a large closed-gate energy barrier for the efflux ion. The needed residue could be on a second loop P2 that folds back into the U2 gating cavity (Fig. 9-2). The residue V474 would be near the critical location A1. Another reason residues on P2 may be folding back into the U2 gating cavity comes from experiments at Gary Yellen's laboratory. It was determined that residues V474, P475, V476 and I477 have a large change in the modification rate constant for reaction with MTSET, when going from a closed to an open channel (Liu et al., 1997; Choi et al., 1993). The modification was for a mutation of each residue to cysteine. Residues beyond I477 would be outside the gated region and towards the cytoplasm. The loop back configuration shown for P2, in Fig. 9-2, is similar to the pore loop P1, except with different residues.

9-6. Coupling between tunnel-track electrons

An important factor influencing gating kinetics is the coupling between adjacent tunnel-track electrons. The repulsive force between the electrons would likely cause distortion of the probability distribution for

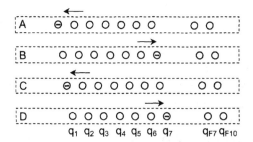

Fig. 9-8. The strong repulsive force between nearby electrons causes pairs of electrons to have a high probability to be at opposite ends of the tunnel tracks. The positive charges in the channel would reduce these repulsive forces, on average; however, a rapid change in the location of an electron, such as occurs when an electron tunnels from q_7 to q_{F7} and then to q_{F10}, could produce a strong transient reduction in the force on electron(s) at the control sites.

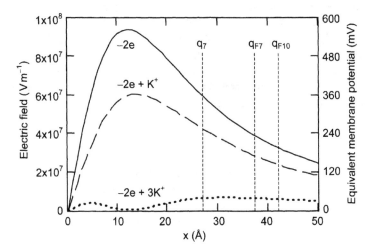

Fig. 9-9. The component of the electric field intensity along the x-axis, at the center-axis of an S4 α-helix, is plotted (solid line) as the distance x from the plane of the q_1 control sites is varied. The field is from an electron at each q_1 control site of the two adjacent subunits, both at $x = 0$. With a single K⁺ ion in the channel at $x = 0$, the electric field is attenuated, but is too large to allow gating in the normal membrane voltage range (dashed line). With two additional ions in the channel, located at plus and minus 5.5 Å from the center ion, gating becomes possible (dotted line). Thus, there must be at least three positive charges in the channel to permit electron gating in the normal membrane voltage range.

the electrons at the tunnel-track sites. It would tend to cause electron pairing at opposite ends of the tracks (Fig. 9-8), to minimize the potential energy of the system. The positive charges in the channel would reduce the coupling and the tendency for electron pairing. Some questions relating to the coupling between tunnel-track electrons are:

1. How much repulsive force is there between the electrons?
2. How much energy is required to move all electrons to the q_1 control sites?
3. How are electrons at far sites influenced by the coupling?

To get a better understanding of these issues, the electric field was calculated ($\varepsilon_r = 4$) for a distance x from the plane of two electron charges at opposite q_1 control site locations, such as tracks A and C in Fig. 9-8. The object here was only to get a rough idea of the interacting forces at various locations across the protein. To simplify calculations, the q_1 control sites were placed in alignment at $x = 0$ and the x-axis was taken as the centerline for the α-helix of track B. The q_1 sites were located 10 Å from the center of the channel and 14.14 Å from each adjacent q_1 site. The curves plotted (Fig.9-9) are for the component of the electric field that is in the direction of the x-axis and parallel to the assumed direction of the membrane electric field. Thus, the field can be represented in terms of an equivalent membrane potential. The following can be inferred from this approximate analysis:

1. At least three positive charges would be required in the channel to reduce the equivalent membrane potential to the level needed for voltage dependent electron gating.
2. Electrons at far sites produce a large electric field at the q_1 control sites of adjacent tunnel tracks – about 3×10^7 Vm^{-1} for q_{F10}. These fields may be reduced by positive charges in the channel (such as in or near the U2 cavity), but rapid changes in the electric field from electron tunneling might be reflected back to the q_1 control sites without much attenuation on a short time scale. Thus, the coupling between tunnel track electrons and the tunneling time constants are likely to play a role in both the activation and inactivation kinetics of electron-gated ion channels.

9-7. Kinetics and inactivation depend on far sites

With protein sequences for many channels now available, it is interesting to compare the channel characteristics with the location of the far sites. The potassium channels have a wide variety of inactivation time constants, which are determined mainly by the location of the controlling far sites and the inactivation-gate cavities. Only the fast millisecond inactivation associated with the U3* cavity does not use a far site. In *Shaker* this cavity is closed by one or two residues on the N-terminus chain. The chain is inserted into the interior mouth of the channel, with the residues aligned to close the cavity. This is inferred from the electron-gating model and is consistent with mutation experiments (Hoshi et al., 1991; Zagotta et al., 1990; Hoshi et al., 1990). Four N-terminus chains are likely inserted, one for each potassium subunit. The ions then pass sequentially through four closed cavities. This would allow inactivation by an electron in any one of the four tunnel tracks. If only one N-terminus chain was inserted, then only one of the four tunnel-track electrons could cause fast inactivation. The Shab K_V2 channels have far sites, but are reported not to inactivate. The K_V2 far site has a $q_5 - q_{F13}$ tunneling distance of 20.9 Å and a peak time constant of ~1 hour at 6°C. The channel may inactivate on this time scale.

Kinetics of both activation and inactivation are influenced by the far site locations because of the interacting force between the electrons. Ions and toxins entering the internal mouth of the channel can influence electron probability at the adjacent far sites and this change in electron probability could then couple back and influence the activation gating. The ion channel would function as an interacting electrostatic system with coupling between adjacent tunnel-track electrons and between the tunnel-track electrons and ions or other charges in the channel, or just outside the channel. Thus, far-site locations play an important role in determining the kinetics of channel gating.

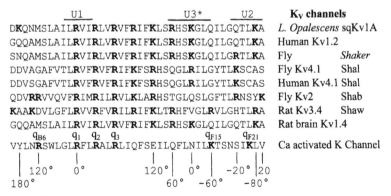

Fig. 9-10. Far-site, near-site, and back-site locations on the S4 transmembrane protein segments illustrate the diversity for several categories of potassium channels. All of the categories have differing far-site locations. The far-site locations determine inactivation-gating kinetics, but can also affect activation gating kinetics by the interaction between electrons in adjacent tunnel tracks. Gating cavities U1, U2 and U3* are located on S6. Inactivation using cavity U3* requires residues from the N-terminus region to be present.

PART II

EXPERIMENTAL
MICROWAVE INVESTIGATION

Chapter 10

MICROWAVE THERMAL FLUORESCENCE SPECTROSCOPY

10-1. Microwave spectroscopy for caged proteins

After developing a molecular stretching model for amplification (Chapter 2) and calculating a frequency of 21 GHz, near the inversion frequency for NH_3, it became an objective to experimentally verify the calculated frequency. This would also help to verify the amplification hypothesis. Microwave spectroscopy is normally restricted to the gas phase because strong water absorption bands mask the sample absorbance in the liquid phase. Could there be a way around this problem? What if fluorescence was used instead?

When microwaves are absorbed by a sample, there is a temperature rise. Most fluorescent molecules have a small temperature coefficient and I wondered if, by using temperature-sensitive fluorescence detection (optimized to measure a small signal on top of a big one), the microwave absorption of a sample could be detected, with less interference from water. To detect a temperature change of the arginine and lysine NH_3 side chain groups, the fluorophore would need to be close to these residues and protected from intimate contact with the surrounding water molecules.

To conduct the experiment, a new type of spectroscopy instrument for recording protein spectra using thermally modulated fluorescence, would have to be constructed and a microwave swept signal generator and amplifier would need to be obtained. It all seemed like a long shot, but I thought it would be worth trying.

BioKinetix had been manufacturing a chemistry analyzer for clinical diagnostics. One of the features of the analyzer was a sensitive detection system, capable of detecting very small changes in photometric absorbance. This low-noise technology permitted measurement of very low enzyme activity, which was advantageous for the early detection of the heart isoenzyme CK-MB by kinetic analysis. The analyzer was used in hospitals for monitoring cardiac patients.

I decided to adapt the chemistry analyzer technology to measure fluorescence in a new type of spectroscopy instrument. The necessary microwave plumbing was installed to irradiate the sample with microwave energy and a laser diode was added to excite fluorescence. A laser diode was the ideal source because it is easy to regulate the output intensity to a few parts per million and it has a narrow emission beam and a relatively narrow bandwidth. The laser diode used was the NSHU590, with a peak emission at 375 nm and 1 mW output. This was the first commercially available UV laser diode using gallium nitride, a new UV/blue light emitting technology, pioneered by Nichia Chemical and its now famous inventor Shuji Nakamura.

Microwave Thermal Fluorescence (MTF) Spectroscopy was thought to be an appropriate name for the spectroscopy technique. The instrument configuration (Fig. 10-1) used a synthesized swept signal generator and a separate microwave amplifier that covered the range up to 26.5 GHz and was capable of a maximum output of about 400 mW. A one-quarter wavelength ground plane antenna, consisting of a 5 mm radiating extension from a semi-rigid microwave coaxial cable, was located about 4 mm from the sample. The 5 mm radiating element, extending vertically above an aluminum bottom plate, gave a close impedance match to the 50-ohm coaxial cable in the region of 15 GHz. An external pulse generator modulated the swept signal generator with a 1 Hz square wave, producing a pulsed RF output. This pulsing caused the local temperature at the sample absorption sites to cycle and the fluorescent emission to cycle with the temperature. The absorbed power resulted in overall heating of the sample, which caused a drift in the fluorescence intensity.

The amplitude of the recorded spectra (Fig. 10-2) is proportional to the average time derivative of fluorescence intensity. The amplitude is also proportional to the fluorophore temperature change ΔT_F, which varies with the absorbed microwave power. Spectrum scans were made using a differentiator/filter to remove the temperature drift signal. This left the higher frequency signal, which caused the recording trace to follow the

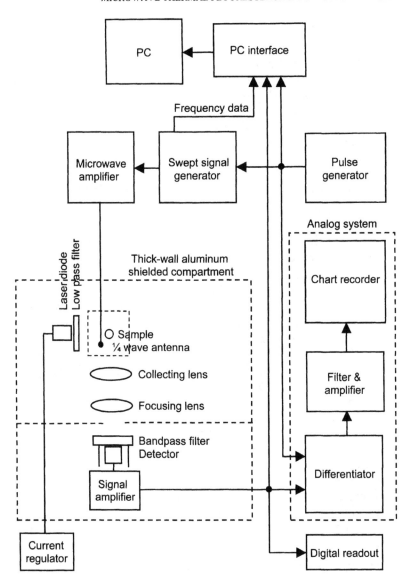

Fig. 10-1. Block diagram of the Microwave Thermal Fluorescence Spectrometer. As the frequency of the pulse-modulated microwave signal is scanned, an oscillating temperature can occur at the sample absorption sites. The heat is conducted to nearby fluorophores, causing an oscillating fluorescence emission. The oscillating fluorescence is detected and sent to a differentiator where the fluorescence drift from overall sample heating is removed. The amplitude for the recorded spectra is proportional to the change (ΔT_F) in fluorophore temperature, which varies with the absorbed microwave power.

1 Hz cycling of the fluorescence intensity. A chart recorder, with a time base of 1 cm per minute, recorded spectra over the range of 8 GHz to 26.5 GHz. The scanning rate was 10 MHz per second. This gave a scale of 600 MHz/cm on the chart recordings. As shown in Fig. 10-2, there are strong water absorption bands in the region between 9 GHz and 11.3 GHz. Above 11.3 GHz, the water absorption is attenuated. The recordings in Fig. 10-2 and Fig. 10-3, for the BFP samples, were made with a nominal 50 mW power output into the antenna. For the two blank recordings with blue fluorescent microspheres, the power output was about 125 mW.

10-2. Microwave spectra for Blue Fluorescent Protein

The green-fluorescent protein of the jellyfish *Aequorea victoria* consists of a single chain of 238 amino acids. It has a natural fluorescence at 508 nm when excited at its absorption maxima of 395 nm. The protein chain forms an 11-stranded β-barrel wrapped around a single central α-helix. The barrel (referred to as a β-can) forms a near perfect cylinder about 42 Å long and 24 Å in diameter that shields the internal fluorophore from the external solvent (Ormö et al., 1996). The fluorophore is near the center of the α-helix and consists of residues Ser-65, Try-66 and Gly-67.

The samples used in this study were Blue Fluorescent Protein (GFP-Y66H) from a control material by Qbiogene, Carlsbad, California (#AFP-5102). These samples differed from the wild type GFP by three residues. A mutation, Y66H shifted the emission peak to the blue region at 450 nm and had an excitation peak at 387 nm, which was close to the laser diode emission peak at 375 nm. The two other mutations (F64L and V163A) were for improved solubility and protein folding at 37°C. The GFP protein has 7 arginine and 21 lysine amino acids. Arginine R96 is only 2.7 Å away from the Try-66 site of fluorescence (Yang et al., 1996). The structure of GFP suggested that some arginine and lysine residues were protected from the solvent and that their NH_3 side chain groups could be undergoing inversion resonance. Microwave absorbance spectra for the NH_3 inversion resonances might then be detectable as a temperature change at the fluorophore.

The sampling method and sample blanks

At first, 3 mm square fluorescence cuvettes were used to hold the BFP samples, and samples were diluted 1:4 with Tris buffer. However, after some experimenting it was decided to use the BFP control in undiluted form, by aspirating 10 µL into the tip of 100 µL glass disposable micro-sampling pipettes. This was the sampling method used for the recordings shown in Fig. 10-2 and Fig. 10-3.

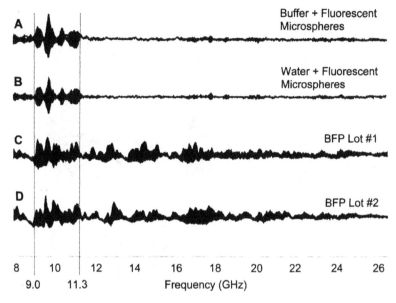

Fig. 10-2. Microwave spectra for Blue Fluorescent Protein taken from two different lots and for two sample blanks containing Blue Fluorescent Microspheres.

The spectrum of 8 GHz to 26.5 GHz was scanned for two sample blanks (Fig. 10-2*A,B*). One blank contained Blue Fluorescent Microspheres (0.5 μm, Duke Scientific #B500) diluted with distilled water. The other blank was diluted with 20 mM potassium phosphate buffer.

10-3. Matching frequencies

The frequency scans for BFP showed numerous peaks in the 11.8 to 20 GHz region that were not due to water. Nearly all the recorded peaks in this region matched the strongest NH_3 gas-phase inversion lines, scaled down with two different scaling factors. These two sets of scaled frequencies were thought to be for two NH_3 groups on the arginine side chain. The forces from the rotations, designated by J,K quantum states, alter the effective tunneling distance for inversion and produce a family of inversion frequencies. The inversion frequency with the maximum intensity was at 23.87 GHz, which was scaled to 16.83 GHz for the arginine NH_3 Group-1 and to 14.39 GHz for arginine Group-2.

Absorption lines in the recorded spectra for BFP (Fig. 10-3) are typically several hundred MHz in width. In the gas phase, the widths of NH_3 absorption lines are strongly dependent on the gas pressure.

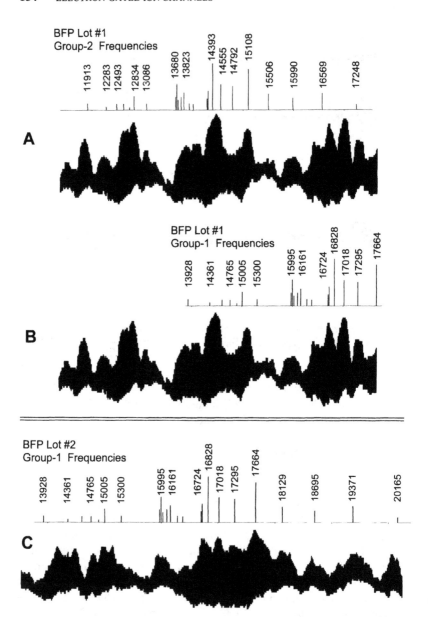

Fig. 10-3. Alignment of scaled gas phase NH$_3$ inversion lines, from Table 10-1, with the recorded microwave spectra for Blue Fluorescent Protein.

Table 10-1. Microwave inversion lines for $N^{14}H_3$

Rotational State J,K	Intensity cm^{-1}	f_{GP} MHz	f_{G1} MHz	In spectra of Lot #2	f_{G2} MHz	In spectra of Lot #1
Observed $N^{14}H_3$ inversion lines (in gas phase) (Townes and Schawlow, 1975)			Group-1 Scaled frequencies $f_{G1} = f_{GP}$ x 0.705		Group-2 Scaled frequencies $f_{G2} = f_{GP}$ x 0.603	
6,3	1.1 x 10^{-4}	19,757	13,928		11,913	x
5,2	5.6 x 10^{-5}	20,371	14,361	x	12,283	x
8,6	1.0 x 10^{-4}	20,719	14,606		12,493	x
6,4	9.9 x 10^{-5}	20,944	14,765	x	12,629	x
4,1	4.0 x 10^{-5}	21,134	14,899		12,743	x
5,3	2.3 x 10^{-4}	21,285	15,005	x	12,834	x
4,2	1.1 x 10^{-4}	21,703	15,300		13,086	x
5,4	2.2 x 10^{-4}	22,653	15,970	x	13,659	x
4,3	4.4 x 10^{-4}	22,688	15,995	x	13,680	x
6,5	1.7 x 10^{-4}	22,732	16,026	x	13,707	x
3,2	2.2 x 10^{-4}	22,834	16,097	x	13,768	x
7,6	2.9 x 10^{-4}	22,924	16,161	x	13,823	x
2,1	1.1 x 10^{-4}	23,098	16,284		13,928	x
8,7	9.9 x 10^{-5}	23,232	16,378	x	14,008	x
1,1	1.9 x 10^{-4}	23,694	16,704	x	14,287	x
2,2	3.2 x 10^{-4}	23,722	16,724	x	14,304	x
3,3	7.9 x 10^{-4}	23,870	16,828	x	14,393	x
4,4	4.3 x 10^{-4}	24,139	17,018	x	14,555	x
5,5	4.0 x 10^{-4}	24,532	17,295	x	14,792	x
6,6	6.9 x 10^{-4}	25,056	17,664	x	15,108	x
7,7	2.7 x 10^{-4}	25,715	18,129	x	15,506	x
8,8	2.0 x 10^{-4}	26,518	18,695	x	15,990	x
9,9	2.8 x 10^{-4}	27,478	19,371	x	16,569	x
10,10	9.0 x 10^{-5}	28,604	20,165	x	17,248	x

Maximum intensity (3,3): f_{GP} = 23.870 GHz f_{G1} = 16.83 GHz f_{G2} = 14.39 GHz
Inversion frequency (0,0): $f_{0\text{-}GP}$ = 23.786 GHz $f_{0\text{-}G1}$ = 16.8 GHz $f_{0\text{-}G2}$ = 14.3 GHz

At a gas pressure of 0.83 mm Hg, the one-half amplitude width for the NH$_3$ 3,3 line was 50 MHz (Townes and Schawlow, 1975) due to pressure broadening from collisions with other gas molecules. At a pressure of 0.27 mm of Hg, the absorption line had a width of 16 MHz. At very low gas pressure the absorption line width can be less than 1 MHz. The recorded half-amplitude line widths of ~200 MHz, for the NH$_3$ lines in GFP-Y66H would correspond to a gas pressure of about 3.4 mm Hg; however, calculations (Eq. 10.9) indicate that there are no gas molecules within the GFP cage. The broadening of NH$_3$ absorption lines therefore must be due to thermal vibrations of the side chain.

The calibration lines for the scaled frequencies, shown in Fig. 10-3, were generated by a graph-plotting program that used data from Table 10-1 for frequency f_{GP} and the intensity data for the line height. The table includes all frequencies with an intensity of 9×10^{-5} or greater (Townes and Schawlow, 1975). Scaling the frequency f_{GP} by 0.603 (Group-2) gave alignment of the strongest gas-phase lines with the strongest recorded peaks in the 14 GHz region. Virtually all of these scaled frequencies gave close alignment with the BFP recorded absorption peaks. However, some of the recorded peaks in the 16 GHz region could not be accounted for. A second series of frequencies was calculated by scaling f_{GP} with the factor 0.705 (Group-1). This factor gave lines having alignment with the remaining unaccounted-for BFP peaks in the 16 GHz region. One BFP control, from a different lot (Lot #2), showed strong spectra for Group-1, but had weak spectra for Group-2. The Group-1 frequencies showed good alignment with the BFP spectra in the region of 16 GHz and above. These two sets of scaled frequencies can account for virtually all of the recorded BFP spectra in the region of 11.8 to 20 GHz.

The unexpected finding of two sets of scaled frequencies led to the hypothesis that arginine has two NH_3 side chain groups undergoing inversion resonance; one at 16.8 GHz, with a single bond between nitrogen and carbon, and the other at 14.3 GHz with a more restrictive double bond to carbon. The lysine NH_3 group, having a single bond to carbon, would have the higher frequency. For the single bond, calculations for the reduced mass μ (Section 2-8) indicated that adding the mass of one carbon atom to the nitrogen mass, would give (approximately) the observed Group-1 frequency reduction. For the double bond, adding to the calculation the mass of a carbon atom and a second nitrogen atom gave reasonable agreement with the Group-2 frequency reduction. Since a double bond restricts rotation about the bond, the rotations for the Group-2 frequencies may be about a single bond between the carbon atom and the second nitrogen. The double bond to carbon apparently causes the inverting NH_3 group to be influenced by the mass of both the carbon atom and second nitrogen atom.

10-4. Estimating parameters and sensitivity

The microwave spectra indicated that the caged BFP fluorophore has a very high sensitivity to microwave energy transitions in nearby arginine amino acids. For the samples in this study, there were 3×10^6 more water molecules than BFP molecules in the sample, yet the fluorescence peaks for the amino acid energy transitions and the peaks for the water molecules had about the same amplitude. This surprising discovery, which makes possible MTF spectroscopy, will be examined in this section.

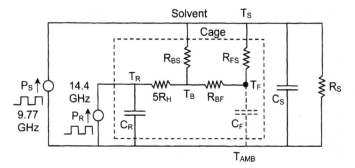

Fig. 10-4. The BFP caged fluorophores in this study have a remarkably high sensitivity to microwave energy transitions in nearby arginine amino acids. This sensitivity is illustrated by the thermal fluorescence scans for blue fluorescent protein. The high sensitivity is needed for detecting the microwave transitions because there are many orders of magnitude more solvent molecules than arginine and lysine absorbers. The circuit and parameter values show the thermal characteristics associated with the caged fluorophore that make this high sensitivity possible. The temperature change at the fluorophore ΔT_F is likely dominated by the microwave absorption of the nearest arginine residue. Residue R73 is the nearest in the amino acid sequence (Fig. 10-6*A*).

Temperature coefficients

Temperature coefficients were measured for the fluorescence of BFP, for the sample blanks with Blue Fluorescent Microspheres, and also for NADH, which had been used as a test sample. The following temperature coefficients were recorded:

$-0.034\ ^\circ C^{-1}$ BFP
$-0.023\ ^\circ C^{-1}$ NADH solution
$-0.016\ ^\circ C^{-1}$ Blue Fluorescent Microspheres in water

All of the chromophores were excited at 375 nm and the fluorescence was monitored at 450 nm using a filter bandwidth of 50 nm. The temperature coefficients were measured using a 3 mm square fluorescence cuvette with a thermocouple placed into the sample and buffer solution for recording temperature. The solution was heated by microwave absorption at 9.77 GHz, a region of maximum water absorption. The temperature rise of the solution was recorded versus time, while the intensity of the fluorescence signal was monitored on a digital readout.

Using NADH as a test sample, the system noise was 7×10^{-5} for the fluorescence intensity with the microwave power off. This increased to 1.8×10^{-4} off-peak baseline noise with the power on, corresponding to a $8 \times 10^{-3}\ ^\circ C$ peak-to-peak temperature fluctuation. In the recordings of

Fig. 10-2C and Fig. 10-3A, the peak-to-peak temperature for the highest peak at 14.4 GHz was 0.12°C and for the highest water peak at 9.77 GHz it was 0.16°C. The baseline noise level for a water sample blank was about 0.01°C peak-to-peak over most of the range above 11.3 GHz.

Thermal parameters for a BFP sample

In the circuit analogy of Fig. 10-4, power corresponds to current flow and temperature corresponds to voltage. The capacitors store thermal energy. They are charged by the pulsed microwave power absorbed by the amino acids (P_R) or absorbed by the solvent (P_S). C_S is the thermal capacitance for the solvent. For the experiments, the total sample volume was 10 µL, most of which was water. Using the heat capacity for water, and 10 µl for the water-solvent volume, the thermal capacitance C_S was estimated as

$$C_S = (4.2 \text{ Jg}^{-1}\,^{\circ}\text{C}^{-1})(10^{-3} \text{ g}\,\mu\text{L}^{-1})(10\,\mu\text{L}) = 0.042 \text{ W s}\,^{\circ}\text{C}^{-1}. \quad (10.1)$$

The thermal time constant for a temperature rise of the solvent was estimated from a recording of BFP fluorescence, as the solvent was heated by the water absorption at 9.77 GHz. Coupling of the solvent to the glass pipette and then to the ambient can be represented, on a fast time scale, by a time constant $\tau_S = R_S C_S$ of about 12 seconds. On a longer time scale, there is a second time constant τ_{GA} of about one minute for coupling the glass to the ambient. From this, the thermal resistance for the fast time constant is given by

$$R_S = \tau_S / C_S = (12\text{s})/(0.042 \text{ W s}\,^{\circ}\text{C}^{-1}) = 280 \,^{\circ}\text{C W}^{-1}. \quad (10.2)$$

All amino acids (except proline) have an identical backbone molecular group. After forming a peptide bond between amino acids and releasing a water molecule, the formula weight for the backbone group is 56. The amino acid glycine has only a hydrogen atom in the side chain. It has a formula weight of 75 and a heat capacity of 99.2 $\text{JK}^{-1}\text{mol}^{-1}$ (NIST data at 25°C). The thermal capacity C_B for the backbone group should be close to that of the glycine, scaled down by the ratio of the two formula weights.

$$C_B = (99.2 \text{ J}^{\circ}\text{K}^{-1}\text{mol}^{-1})(56/75) = 74 \text{ W s}\,^{\circ}\text{C}^{-1}\text{mol}^{-1} \quad (10.3)$$

The heat capacity for the gas NH_3 at 25°C and 1 bar is 35.0 $\text{JK}^{-1}\text{mol}^{-1}$ (NIST data).

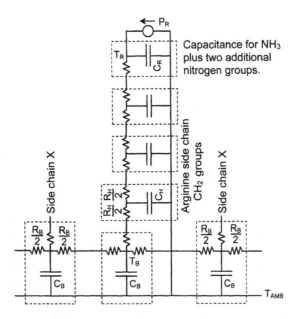

Fig. 10-5. In the thermal model, microwave power P_R is absorbed by an ensemble of arginine NH_3 groups. This absorbed power charges the ensemble thermal capacitance C_R and produces a temperature rise ΔT_R. The increase in temperature causes heat to be conducted from C_R through the side chain resistance $5R_H$ and backbone network, charging the capacitors and producing a temperature rise ΔT_F at the fluorophore location. An additional resistance R_H is included in the $5R_H$ to account for the coupling between the nitrogen groups. Shunting resistances and heat loss to the solvent (Fig. 10-6) attenuate the temperature rise at the fluorophore and at the NH_3 absorption site. The BFP cage substantially increases the thermal resistance to the solvent and thus reduces the heat loss.

For the amino acid arginine, an approximate thermal circuit (Fig. 10-4) has a thermal capacitance C_R, representing the energy stored by microwave absorption in the NH_3 groups. Arginine has three nitrogen groups at the end of its side-chain, which are connected to the backbone by a chain of three CH_2 groups (Fig. 10-5). The thermal capacity for the three nitrogen groups was estimated as

$$C_R = 3(35.0 \text{ J}^\circ\text{K}^{-1}\text{mol}^{-1}) = 105 \text{ W s} ^\circ\text{C}^{-1}\text{mol}^{-1}. \qquad (10.4)$$

The heat transfer from capacitance C_R to the backbone would be across the thermal resistance $5R_H$ (an additional R_H was included for N coupling).

The relevant experimental observations that established constraints for the circuit-model parameters were:

1. There is approximately a 10-second charging and discharging time constant for the temperature change at the fluorophore in response to a step change in power at the NH_3 resonant microwave frequencies.

2. When a large step change in power was applied at the frequency of a water absorption peak (9.77 GHz), the delay in the temperature rise at the fluorophore (transport lag) was less than 1-second.

3. The response to the 1 Hz square wave pulse of microwave power was roughly that of a triangular pulse, after the low frequency temperature drift was removed.

The circuit in Fig. 10-6B shows the heat conduction path for several NH_3 absorption sites. The limiting time constant is for capacitance C_R times the total resistances for heat flow to the solvent. This is the only path that could account for the observed 10-second time constant at the fluorophore (constraint #1) and not introduce a long transport lag for a step change in power at the the 9.77 GHz frequency (constraint #2). Most of the thermal resistance for the time constant is from the long arginine or lysine side chain. The approximate locations for resistance paths to the solvent are shown in Figure 10-6A. They are based on the protein sequence and the β-can geometry (Ormö et al., 1996; Yang et al., 1996) for GFP.

A simplified circuit for the arginine site nearest the fluorophore (in the protein sequence) is shown in Fig. 10-6C. In order to make calculations, some assumptions need to be made for the relationship between the various thermal resistances. The assumptions are:

1. Resistances R_B and R_W are equal to R_H

2. Resistance R_B' is equal to one-fifth R_B because of cross bonding

In an α-helix, the shape is maintained by hydrogen bonding between every third residue and this decreases the thermal resistance across the α-helix. The thermal resistances near the fluorophore are for an α-helix and are designated as R_B' to account for a lowered thermal resistance because of cross bonding of the α-helix.

From the thermal circuit (Fig. 10-6C) and the assumptions listed above, a temperature rise ratio was calculated for the absorption site R73.

$$\Delta T_R/T_F = 7.4 \qquad (10.5)$$

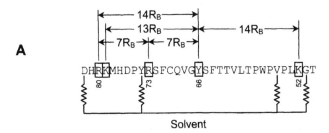

A

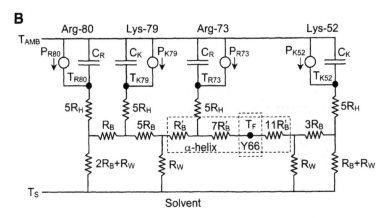

B

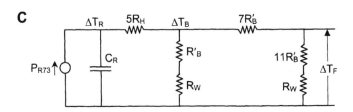

C

Fig. 10-6. The amino acid sequence and the β-can geometry for GFP indicate the likely locations for thermal resistance to the solvent (Fig. A). The shunting resistances to the solvent, near the backbone group(s), cause most of the heat flow to bypass the fluorophore. The shunting resistance limits the temperature rise at the NH_3 absorption site and creates a voltage divider type of attenuation for ΔT_F at the fluorophore. Fig. B shows a thermal circuit for heat flow to the fluorophore from nearby arginine and lysine sites. From Fig. B, it follows that most of the temperature rise at the fluorophore would be from photons absorbed by Arg-73. A simplified circuit for this thermal path is shown in Fig. C.

Also from Fig. 10-6C, the time constant for a fluorophore temperature rise, resulting from a step change in power P_R, would be approximately

$$\tau_R = 6 R_H C_R. \qquad (10.6)$$

If we assume a time constant of 5-seconds and the above assumption $R_B = R_H$, then an estimate for thermal resistance R_B would be

$$R_B = \tau_R / 6 C_R = (5\,\mathrm{s})/(6 \times 105\,\mathrm{W\,s\,°C^{-1}mol^{-1}}) = 8 \times 10^{-3}\,\mathrm{°C\,W^{-1}mol}. \quad (10.7)$$

From the assumption $R'_B = (1/5) R_B$, a time-constant τ_B can be estimated.

$$\tau_B = R'_B C_B = (1.6 \times 10^{-3}\,\mathrm{°C\,W^{-1}mol})(74\,\mathrm{W\,s\,°C^{-1}mol^{-1}}) = 0.1\,\mathrm{s} \qquad (10.8)$$

The heat flow for a change in the solvent temperature is conducted across a series of backbone groups of the caged α-helix to the fluorophore. Each group adds to the delay. The 0.1 second time constant for a cross-linked backbone sub-group is probably compatible with the observation that the transport lag to the fluorophore was less than one second, for a step increase in microwave power at the 9.77 GHz water absorption peak.

The time constant τ_R of 5-seconds is about one-half of the value listed for constraint #1. If we had considered the additional capacitance values C_H in the arginine side chain, this would have increased the time constant.

Other means of heat transfer

The arginine residue R96 in BFP is reported to be only 2.7 Å from the Y66 fluorophore (Yang et al., 1996). One might think there could be some heat transfer by convection. Calculations for the density of air molecules in the cage indicate that this is not likely. The density of air at one atmosphere and 25°C is 0.00118 g cm^{-3}. Using this and the formula weight for N_2, the calculated number of molecules present in a volume of 1000 Å^3 is

$$N = (0.00118\,\mathrm{g\,cm^{-3}})(1/28\,\mathrm{mol\,g^{-1}})(6.02 \times 10^{23}\,\mathrm{mol^{-1}})(10^{-21}\,\mathrm{cm^3\,nm^{-3}}).$$

$$N = 0.025\,\mathrm{molecules\,nm^{-3}} \qquad (10.9)$$

Thus, in the internal volume of the BFP cage, air molecules are not likely to be present and heat transfer by convection can be disregarded.

It has been suggested that hydrogen bonds from the carbonyl of Thr-62 and from Gln-182 may help to stabilize Arg-96 (Ormö et al., 1996). If this is the case, then there could be some conducted heat transfer from Arg-96 to the fluorophore. This would depend on the orientation of the side chain. An orientation towards the solvent would likely quench any inversion and photon absorption. If the side chain is pointing into the cage, some heat transfer could occur. With the Arg-96 backbone group close to the solvent, the attenuation could be similar to the calculated values for R73.

Absorption Sensitivity

BFP has a molecular mass of 27,000 daltons, and the total amount of BFP in the 10 μL sample was

$$\text{Total BFP} = (0.5\ \mu\text{g}/\mu\text{L})(10\ \mu\text{L})/(27{,}000\ \text{g/mol}) = 185\ \text{pmol.} \quad (10.10)$$

When microwave power is applied, with both the pulse generator and the differentiator disabled, there was initially a linear decrease in the fluorescence, corresponding to a rise in the temperature at the fluorophore. The linear temperature rise was modeled as a single capacitance, integrating the power from photon absorption. In Figs. 10-4 and 10-6, the integrating capacitance for microwave absorption by the solvent is C_S and by Arg-73 it is C_R. With the pulse generator activated, it follows that:

1) On a seconds time-scale, absorbed microwave power P_R is integrated by the capacitance C_R, producing a temperature ramp. The temperature rise after charging by a microwave power pulse is

$$\Delta T_R = P_R \int_0^t \frac{1}{C_R}\, dt \,. \quad (10.11)$$

2) From this, the power absorbed by the dominant arginine ensemble R73 is given by

$$P_R = \frac{C_R \Delta T_R}{t} = \frac{C_R}{t}\left(\frac{\Delta T_R}{\Delta T_F}\right)\Delta T_F\,, \quad (10.12)$$

where $\Delta T_R/\Delta T_F$ is the temperature rise ratio (Eq. 10.5). At the 14.4 GHz peak absorption, the observed temperature rise for the one-half second charging pulse was 0.12°C. If three-fourths of the rise is for Arg-73, then the power absorbed by the dominant Arg-73 ensemble is calculated as

$$P_R = (105 \, \text{W s} \, ^\circ\text{C}^{-1}\text{mol}^{-1})(185 \times 10^{-12} \, \text{mol})(7.4)(3/4)(0.12^\circ\text{C})/(0.5\,\text{s})$$

$$P_R = 2.6 \times 10^{-8} \, \text{W at 14.4 GHz.} \tag{10.13}$$

Using the thermal capacity C_S from Eq. 10.1, the power absorbed by the solvent is

$$P_S = C_S \Delta T_S t^{-1} = (0.042 \, \text{W s} \, ^\circ\text{C}^{-1})(0.16^\circ\text{C})/(0.5\,\text{s})$$

$$P_S = 13.4 \times 10^{-3} \, \text{W at 9.77 GHz} \tag{10.14}$$

$$P_S = 1.0 \times 10^{-3} \, \text{W at 14.4 GHz}$$

The ratio P_S/P_R gives the fluorescence sensitivity enhancement of the caged BFP molecule for the microwave absorption (mainly) at Arg-73. This large amount of enhancement requires a high thermal resistance to the solvent, which would be provided by the BFP cage.

According to the Beer-Lambert law, the photometric absorbance A of a substance is proportional to concentration c and the path length l, and can be expressed as

$$A = \varepsilon_\lambda c l = \log\left(\frac{P_I}{P_T}\right), \tag{10.15}$$

where P_I is incident power and P_T is transmitted power. Here we are interested in the absorbed power, which is $P_A = P_I - P_T$. Rearranging terms gives

$$\frac{P_A}{P_I} = 1 - 10^{-A} = 1 - e^{-2.3A}. \tag{10.16}$$

The penetration depth is the path length l that reduces the transmitted power to the fraction $1/e$ of the incident power. For this length $A = 0.434$.

The graph in Fig. 10-7 shows the absorbance and power ratios for the solvent (mostly water) and for a single arginine residue ensemble, which is likely to dominate the temperature rise at the fluorophore of BFP. The high thermal isolation from the solvent permits detection of arginine microwave absorbance more than four orders of magnitude below the solvent absorbance at the same frequency. Taking the solvent absorbance at 14.4 GHz (point B) as the zero reference, the minimum detectability for NH_3 absorption is about 5×10^{-5} A (point C). This is comparable to the best detectability for UV-visible spectroscopy. Thus, Microwave Thermal Fluorescence Spectroscopy can be a remarkably sensitive technique for recording the spectra of caged biological molecules.

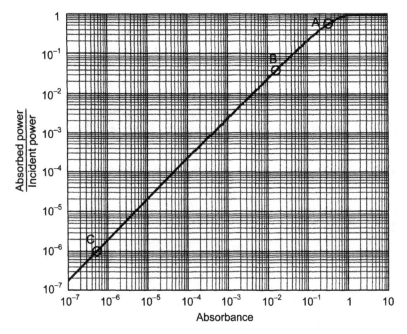

Fig. 10-7. The high sensitivity for detecting microwave absorption in arginine NH_3 (point C) is not impaired by the much greater solvent absorption (point B) at the same frequency. With the thermal isolation of the BFP cage, there is a fluorophore temperature rise from NH_3 photon absorption that is about ten times greater than the solvent temperature rise (B), with less than one-ten-thousandth of the absorbed power. Legend: A) Solvent absorbance at 9.77 GHz peak, B) Solvent absorbance at 14.4 GHz, C) NH_3 absorbance at 14.4 GHz for heating the fluorophore. Incident power: ~25 mW.

10-5. Arginine and lysine hot spots

According to the calculation for the thermal model (Fig. 10-6), the heat from photon absorption at the nearby Arg-73 NH_3 site might cause a temperature rise at the NH_3 site about seven times the temperature rise at the Y66 fluorophore site. However, there probably is some additional heating from at least one other absorption site. If we assume that one-fourth of the temperature rise at Y66 is from another arginine site in the vicinity (Arg-80), then the temperature rise at Arg-73 would be about five times the observed temperature rise for the fluorophore.

The observed temperature rise for the solvent at 14.4 GHz was about one-tenth the fluorophore rise (Fig. 10-2). Thus, the temperature rise at Arg-73 might be ~50 times that of the solvent temperature rise. With 50 mW into the antenna, the observed temperature rise at the fluorophore was 3°C after 20 minutes. Assuming there is no saturation, 200 mW might cause a 12°C rise at the fluorophore and a 60°C average rise at the arginine Arg-73 NH_3 sites (after 20 minutes), with only a 1°C rise in solvent temperature. The calculations depend on the estimated thermal resistance ratio for Fig. 10-6*C*, but a 20 to 60-fold hot-spot temperature rise above the surrounding fluid would seem likely. Other arginine or lysine NH_3 sites could have a similar hot-spot temperature rise, but they would not produce a significant temperature rise at the BFP fluorophore because of their distance and the intervening heat loss to the solvent.

10-6. Calcium oscillators - microwave sensitivity.

Calcium oscillations are known to be sensitive to extremely low frequency (ELF) electromagnetic fields. A frequency of high sensitivity is near 10 Hz. As indicated by the far-site calcium oscillator model (Fig. 8-5), electron tunneling between q_{F5} and q_{F13} corresponds to a frequency of about 10 Hz. and electromagnetic coupling to the electron could be responsible for ELF effects of a calcium oscillator.

Another effect, indicated by the absorption spectra for BFP, is that calcium oscillators may also be sensitive to electromagnetic fields at certain microwave peak absorption frequencies (Fig. 10-2). As indicated above, the NH_3 sites at the end of the arginine side chains can become hot spots when a microwave generator is tuned to the peak absorption frequencies. The NH_3 sites would account for most of the Q_{10} temperature factor in ion channels, since they have the energy barriers for thermally excited electron tunneling. Thus, the hot-spot temperature rise should decrease the electron-tunneling time constant and increase the frequency of calcium oscillations. Tuning a microwave generator to one of the resonant frequencies should selectively increase the far-site rate constants and the calcium oscillator frequency with only a modest rise in the temperature of surrounding fluid. Since calcium oscillators can control neurotransmitter release, prolonged exposure to high radiation levels at these frequencies might produce adverse medical symptoms.

10-7. The first excited vibrational state

The NH_3 ΔE_1 energy value for the first excited vibrational state is of great importance in the electron-gating model, since it is a determining factor in both the amplification and the bandwidth for electron tunneling between adjacent arginine or lysine residues. In the gas phase, ΔE_1 corresponds to a wavenumber of about 36 cm^{-1} and for the ground state it is $(23,786\ MHz)/c = 0.793\ cm^{-1}$.

Scaling these values by the factors listed in Table 10-1 gives the following:

Group-1 $\bar{\nu}_{1\text{-G1}} = 0.705 \times 36 \text{ cm}^{-1} = 25.38 \text{ cm}^{-1} \text{ (760 GHz)}$ (10.17)
$\bar{\nu}_{0\text{-G1}} = 0.705 \times 0.793 \text{ cm}^{-1} = 0.559 \text{ cm}^{-1} \text{ (16.8 GHz)}$

Mode change $\bar{\nu}_{1\text{-G1}} - \bar{\nu}_{0\text{-G1}} = 24.8 \text{ cm}^{-1}$ $(E_{1-} - E_{1+} = 3.09 \text{ meV})$

Group-2 $\bar{\nu}_{1\text{-G2}} = 0.603 \times 36 \text{ cm}^{-1} = 21.7 \text{ cm}^{-1} \text{ (650 GHz)}$ (10.18)
$\bar{\nu}_{0\text{-G2}} = 0.603 \times 0.793 \text{ cm}^{-1} = 0.478 \text{ cm}^{-1} \text{ (14.3 GHz)}$

Mode change $\bar{\nu}_{1\text{-G2}} - \bar{\nu}_{0\text{-G2}} = 21.2 \text{ cm}^{-1}$ $(E_{1-} - E_{1+} = 2.64 \text{ meV})$

The energy difference $E_{1-} - E_{1+}$ is a determining factor for amplification and electron tunneling. Since the microwave spectra for BFP indicated that arginine had both the Group-1 and Group-2 frequencies, it was thought that it might be possible to detect both 21.2 cm^{-1} and 24.8 cm^{-1} as mode changes in the infrared spectra for arginine. These mode changes (according to the amplification model) could occur when an electron tunnels to an arginine site in the presence of a strong electric field.

10-8. Mode switching at infrared frequencies

Time resolved infrared spectroscopy for bacteriorhodopsin photocycle – double-difference spectra show mode switching for arginine R82.

Bacteriorhodopsin is a much studied proton pump that transports a proton across the cell membrane to the cell exterior. This event is triggered by a light photon received at the rhodopsin photoreceptor. An arginine amino acid R82 assists in the transport of the proton across the cell membrane.

Many interesting papers have been published on time-resolved infrared spectra for the bacteriorhodopsin photocycle. One paper that seemed most relevant was on time-resolved infrared spectra for a perturbation of R82 (Hutson et al., 2000). This publication indicates the presence of frequencies attributable to R82 guanidino group vibrations (~1640 and ~1545 cm^{-1}) and suggests that the perturbation may involve a change in the ionization state of R82. An interesting item was on the website of M. Shane Hutson at Duke University for a period of time and was captioned: "Time resolved double-difference spectra of WT–R82A". It showed the infrared double-difference spectra for WT–R82A as the photocycle progressed, after the retinal photoreceptor was excited by a pulsed laser light source. The microsecond time base for the photocycle was scaled down to a seconds time frame, showing shifts in the arginine (WT–R82A) ΔAbs peaks resulting from a perturbation (Fig. 10-8).

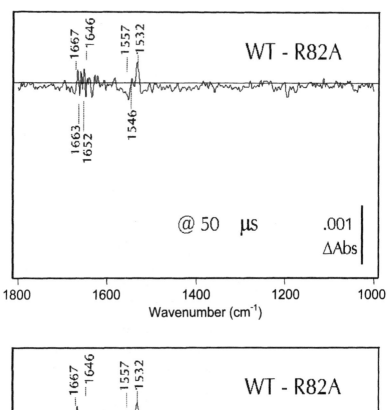

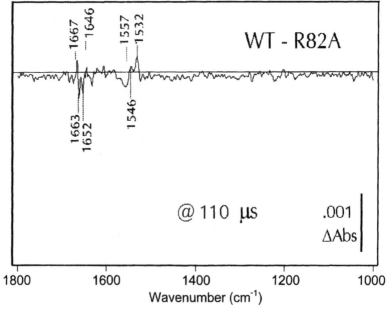

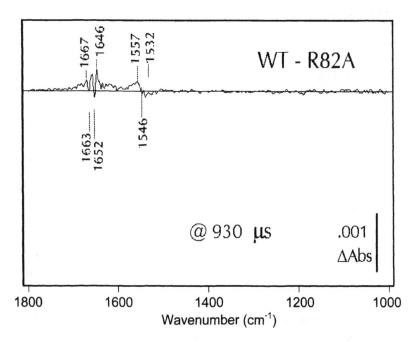

Fig. 10-8. Infrared double-difference spectra for bacteriorhodopsin WT–R82A. The spectra are for the nitrogen groups at the end of the arginine side chain. It shows the spectral changes with time after the rhodopsin photoreceptor is triggered by a pulse of laser light. Stats of the photocycle were taken from the website of M. S. Hutson and are shown here with his permission.

Time-resolved Fourier-transform infrared (FTIR) spectroscopy is an important tool that can show changes in molecular spectra with time on a microsecond time scale, thus providing an insight into reaction mechanisms. With computer processing, rapid frequency transitions can be slowed down for monitoring on a time scale of seconds. The double-difference spectra for WT–R82A is obtained by subtracting the absorbance spectra for bacteriorhodopsin with arginine (R82) replaced by alanine (A), from the absorbance spectra for wild-type bacteriorhodopsin. Thus, the recorded difference spectra (ΔAbs) are only for the arginine (R82) side chain as it responds to a perturbation. The spectra would be for the nitrogen groups at the end of the side chain – just what we are interested in for the electron-gating model.

In observing spectral shifts for the ΔAbs peaks of WT–R82A after triggering of the photocycle, one frequency shift was readily apparent. This was the ΔAbs peak at 1532 cm^{-1} @ 50 μs, which was shifted to 1557 cm^{-1} @ 930 μs, a difference of 25 cm^{-1}. This shift is about the same as the Group-1 mode change of 24.8 cm^{-1}. A second absorbance peak at 1667 cm^{-1} @ 110 μs was shifted to 1646 cm^{-1} @ 930 μs (Fig. 10-8), a difference of 21 cm^{-1}, matching the Group-2 mode change of 21.2 cm^{-1}.

These shifts in the spectral peaks correspond to energy changes matching $E_{1-} - E_{1+}$ of the electron-gating model (Fig. 2-2). This is the energy change for removing a donor electron from an amplifying NH$_3$ arginine site. The energy decrease is accompanied by an NH$_3$ oscillation mode change from antisymmetric to symmetric. This could only occur when the ground state and the first excited state energies are split as a result of the NH$_3$ inversion resonance. It would seem that the infrared frequency shifts must be associated with a change in the oscillating mode of NH$_3$.

Interpretation of ΔAbs peak shifts based on electron-gating model

From the electron-gating model it is inferred that a change from the low-energy mode to the high-energy mode could occur when an electron tunnels to the arginine site and there is a sufficiently strong electric field to cause saturation for the amplified rate curve (Fig. 2-3). The spectra at 50 μs with a peak at 1532 cm^{-1} would represent the arginine NH$_3$ oscillating in the low-energy symmetric mode. This would be for Group-1 with a single bond between the nitrogen and the carbon atom. If an electron tunneled to the NH$_3$ site, the NH$_3$ would then switch to oscillating in the higher energy antisymmetric mode. This would shift the absorbance peak to 1557 cm^{-1}. According to the analysis in Chapter-2 a strong electric field would increase the probability for NH$_3$ with a donor electron to be found in the high-energy mode. With no external electric field the oscillation could be in either mode and there might be peaks at both 1532 and 1557 cm^{-1}. The ΔAbs changes at 1667 cm^{-1} to 1646 cm^{-1} are more difficult to interpret. It was assumed they are for NH$_3$ Group-2 with a double bond to the carbon atom. From a lysine residue (K216) an electron would have a higher probability to tunnel to the NH$_3$ group tuned to the same $E_{1-} - E_{1+}$ energy (Group-1). This could explain the smaller absorbance change for 1667 to 1646 cm^{-1} (NH$_3$ Group-2). Does bacteriorhodopsin use a tunneling electron? The above analysis suggests that this is a possibility.

In conclusion, the shifts of the ΔAbs peaks for arginine WT–R82A agree with the mode change wavenumbers for the electron-gating model, which are determined from the microwave Group-1 and Group-2 inversion frequencies.

APPENDIX

A. Geometric calculations for an α-helix

Amino acid residues are located every 100 degrees on an α-helix. To determine the α-helix rotational angle ϕ for the NH$_3$ tunneling site centers, a donor tunneling site is taken as the reference, with ϕ equal to 0 degrees. The angle advances clockwise by 100 degrees for each increase in the rise number n, but ϕ is expressed as a positive or negative angle with respect to the 0 degree reference. The angle ϕ is given by

$$\phi = 100n - 360 \operatorname{floor}\left[\frac{n + 2 + \operatorname{floor}\left(\dfrac{n}{13}\right)}{4}\right]. \qquad (A.1)$$

The parameter n is the Rise No. shown in Table 8-1. The designation "floor(x)" is a truncation function (in Mathcad) that gives the greatest integer equal to, or less than x. The calculated angle ϕ has a maximum value of ± 180 degrees. This angle was used in calculating the hypotenuse of a right triangle to obtain the tunneling distance r.

$$r(n) = \sqrt{(1.5n)^2 + \left[2r_a \sin\left(\frac{\phi \deg}{2}\right)\right]^2} \qquad (A.2)$$

The right triangle is defined by the two tunneling-site centers and a third point at the intersection of a cross-sectional plane passing through the zero degree reference site and a flux line passing through the second site, parallel to the α-helix axis. One side of the triangle is the total rise distance of $1.5n$. The second side is the cross-sectional distance between the sites as determined from the radius r_a (for the tunneling sites) and the angle ϕ. The radius used for this model is: $r_a = 3.97$ Å. The electric field is assumed to be parallel to the α-helix axis. The angle θ, in Table 8-1, between the electric field and the direction of tunneling is calculated using the arc cosine.

$$\theta(n) = a\cos\left(\frac{1.5n}{r(n)}\right)57.296 \tag{A.3}$$

B. Time constant for a tunneling distance *r*

Calculated peak time constants for amplified electron tunneling between two arginine sites on an α-helix are shown in Table 8-1. In the calculations, tunneling between two sites, separated by a rise distance x_r of 4.5 Å or 6 Å, is treated as a space jump – since there are no intervening crossings of the α-helix. With four or more residues separating donor and acceptor sites ($x_r > 6$), the α-helix approaches or crosses between the two sites, placing non-amplifying amino acids near the tunneling path. The presence of these intervening amino acids increases the tunneling probability and reduces the time constant, requiring compensation in calculating time constant values. To have compensation that matched that of electron-transfer experiments, the term $\beta_0\sqrt{U}$ for low energy tunneling derived from the Schrödinger equation (Eq. 2.1) was replaced with the distance-decay constant β.

$$\beta = \beta_0\sqrt{U} \tag{A.4}$$

A standard equation for estimating electron-transfer rate across proteins is based on a one-dimensional square tunneling barrier (1DSB) model.

$$K_{ET} = K_{max}\exp\left(-\frac{\Delta G + \lambda}{4\lambda kT}\right)\exp\left[-\beta(r - r_0)\right]. \tag{A.5}$$

Inverting Eq. A.5 and then incorporating $h_w/2$ for the amplified peak time constant gives

$$\tau_{ep}(r) = \left(\frac{h_w(r)}{2}\right)\frac{1}{K_{max}}\exp\left(\frac{\Delta G + \lambda}{4\lambda kT}\right)\exp\left[\beta(r - r_0)\right], \qquad (A.6)$$

where $K_{max} = 10^{13}$ s^{-1}, $\beta = 1.26$ Å^{-1}, $\lambda = 1.431$ eV, $\Delta G = 0$ eV, $r_0 = 2.09$ Å.

Electron transfer across an α-helix has a predicted distance-decay constant of 1.26 Å^{-1} (Winkler et al., 1999; Langen et al., 1995). This gives a slope in agreement with published experimental observations and was used as the value for β. The amplification h_w decreases with $1/r$ for a space jump (Eq. 2.31), but it can be shown that for tunneling between an amplifying and a non-amplifying (but otherwise similar) site that the peak time constant would decrease by approximately $1/h_w$ and voltage sensitivity would be substantially reduced over much of the range. Thus, h_w would likely approach unity faster than $1/r$ as the number of intervening residues increases. To account for this, the following equation was used to reduce the amplification with increasing tunneling distance:

$$h_w(r) = (h_w - 1)\left(\frac{r_{min}}{r}\right)\exp\left(-\frac{r - r_{min}}{\lambda_d}\right) + 1, \qquad (A.7)$$

where $r_{min} = 6.0$ Å, $h_w = 25.2$, $\lambda_d = 8$ Å.

Here h_w has a constant value with its distance attenuation term (r_{min}/r) brought outside. The one arbitrary term in the above equations is λ_d, which controls the additional rate of decrease of h_w with increasing r, thereby determining the time constant. Parameter λ_d was set to provide reasonable values for the time constants in the range of $r = 10$ to 16 Å by comparison with experimental data for inactivation time constants. The setting of $\lambda_d = 8$ Å is not very critical since most of the decline is for (r_{min}/r).

Calculations for h_w and τ_{ep} in Table 8-1

The 6 Å and 6.59 Å space-jump time constants were treated as special cases and calculated using Eq. A.6 with $\beta = 2.46$ Å^{-1}, $r_0 = 4$ Å, and $h_w = 25.2$. For the other time constants, Eq. A.7 was first used to determine the value of $h_w(r)$, then calculations were made with Eq. A.6 using the values shown below the equation.

The values used for the compensated time-constant calculations raise a number of questions:

1) Why is the value for β about one-half the space-jump value $\beta_0\sqrt{U}$?

2) Why does r_0 equal 2.09 Å instead of 4 Å?

With the reduction in β, a value of $r_0 = 2.09$ Å is needed to have calibration at the 6 Å space jump distance, but how can this be justified?

One explanation:

If donor and acceptor sites are widely separated and a single intervening site is moved across the distance between them, then the maximum tunneling rate will occur when the intervening site is at a (center-to-center) distance $r/2$ from each end site. This would give an edge-to-edge tunneling distance of $r/2 - r_0$. However, the intervening sites are located around the circumference of the α-helix and the maximum rate would therefore be limited by the longest tunneling distance to an off-center site. There would also likely be a probability for tunneling to several off-center sites. Thus, the increase in the limiting tunneling distance would require that r_0 be less than the theoretical space-jump value of r_0 used with $\beta_0\sqrt{U}$.

Suppose the distance between the two end sites is 12 Å. Then the $r/2$ distance is 6 Å. but given the 4 Å radii to the tunneling sites on the α-helix and the location of the sites, the limiting mean tunneling distance could easily be 8.955 Å. The 2.955 Å increase would show up as a decrease in r_0.

$$\frac{r - r_0}{2} = \frac{12 - 2(4 - 2.955)}{2} \tag{A.8}$$

Thus, calibration of Eq. A.6 with $r_0 = 2.09$ Å is justified for $r = 12$ Å in addition to the $r = 6$ Å distance.

Based on the above, we might have used a theoretical value $(\beta_0/2)\sqrt{U}$ in place of $\beta = 1.26$ Å^{-1} for Eq. A.6. This calibration would give peak time constants about one-half the value listed in Table 8-1 for distances near 27 Å and about 15% less than the listed values in the 10 to12 Å range.

FINAL COMMENTS

This book was written to document the findings of a five-year intensive study of ion-channel gating, amplification and electron tunneling. Much information was obtained from a detailed examination of the Hodgkin-Huxley equations as they were compared to a model for electron gating. In developing the gating model it was initially assumed that amplification was constant. This worked well in the normal membrane voltage range; however, it became apparent that the amplification would be reduced to unity and the rate curves would saturate for a large tunneling voltage. Finite-range rate constants were then developed, which caused the rate curves to merge with a curve for the Marcus inverted region when the amplification decreased to unity. Another model for the finite-range rate constants might use voltage-sensitive amplification, but it was not exactly clear how to accomplish this. After completing most of the book and while reviewing Chapter 2, it was discovered that the NH_3 dipole moment interacting with the donor electron would make the amplification voltage-sensitive. This new understanding of the amplification process led to a voltage-sensitive amplification equation, which eliminated the need for the earlier, and more complicated, equations for finite-range rate constants. These earlier equations have been retained in the book and the agreement between the two is shown graphically in Chapters 2 and 6. Sections 2-6, 2-7, 4-7 and 6-4 of the book were the last sections to be written and incorporate the voltage and energy sensitive amplification.

A brief review of the findings

1. Arginine and lysine amino acids have NH_3 group(s) at the end of their side chain, which cause amplification for electron tunneling across the S4 transmembrane protein segments of ion channels. The amplification is caused by inversion of the NH_3 groups and a mode switching in their oscillation (phase modulation) that alters both the energy and the average displacement of a donor electron in response to a change in the external electric field. (Ch. 2)

2. From microwave experiments with Blue Fluorescent Protein it was determined that there are two inversion frequencies, one at about 14.3 GHz and a second at 16.8 GHz, corresponding to two NH_3 side chain groups on arginine. The inversion frequencies were determined by scaling down the published NH_3 gas-phase rotational-vibrational absorption lines by two separate factors, which were adjusted to produce alignment with recorded spectra for BFP. The two frequencies could be accounted for by increasing the reduced mass of NH_3. Adding the mass of a carbon atom, lowered the NH_3 gas-phase inversion frequency to 17.0 GHz. Adding an additional nitrogen mass gave a further reduction to 14.6 GHz. The two inversion frequency reductions are for a single and a double bond between the nitrogen and the carbon atoms. (Ch. 10, Ch. 2)

3. Infrared double-difference spectra for arginine WT–R82A, in the bacteriorhodopsin photocycle, were recorded with Time Resolved FTIR Spectroscopy in a study by M. Shane Hutson. The spectra showed a peak absorbance shift of 25 cm^{-1}, matching the 24.8 cm^{-1} mode change, associated with the Group-1 inversion frequency for arginine. There was also a smaller shift of 21 cm^{-1}, matching the 21.2 cm^{-1} mode change associated with the arginine Group-2 inversion frequency. These IR spectral shifts, for arginine R82 in bacteriorhodopsin, lend support to the calculations for the electron-gating model and the two inversion frequencies determined for arginine in Blue Fluorescent Protein. (Ch. 10)

4. For the lower inversion frequency the amplification by the NH_3 group is about 25-fold, in the normal ion channel voltage range and with a 6 Å distance between the arginine tunneling sites. The NH_3 group with a donor electron is sensitive to the electric field, so that its amplification factor decreases with increasing tunneling distance. The amplification also decreases to unity for large electric

field intensities. For the higher inversion frequency, which would apply to lysine, the amplification is about 22-fold. (Ch. 2, Ch. 3)

5. Amplification reduces the input energy, absorbed from an external electric field, to transfer electron charge across the tunneling distance. The energy gain and power gain is about 25-fold. (Ch. 2, Ch. 3)

6. Amplification increases the time constant and the dwell time and reduces the energy required for electron transfer. These factors reduce the probability for the tunneling electron to be found at non-amplifying sites, thus resulting in a selective tunneling path across arginine and lysine residues of the protein. (Ch. 4)

7. Using equations for the electron gating model (Table 6-1C) and Eq. 7-5, an amplification of $h_w = 25.2$ gives an I-V curve in agreement with the well-known, experimentally determined, negative-conductance region of the sodium channel. When amplification is reduced to unity, the negative conductance is eliminated. The 25-fold amplification of electron tunneling, when combined with a membrane voltage attenuation factor of $\eta = 0.106$, gives a rate constant with a sensitivity matching the observed sensitivity for the β_m rate constant of the sodium channel. (Ch. 1, Ch. 2, Ch. 6)

8. Gating currents are due to the tunneling of electrons across the amplifying tunneling sites on the S4 transmembrane segments. The gating-current time constants are determined, in part, by the distances between the arginine or lysine residues as established by the amino acid sequence. Charge immobilization is due to tunneling electrons being retained at arginine or lysine far sites on the S4 transmembrane segment. (Ch. 8)

9. The Hodgkin-Huxley equations allowed calculation of the open-gate energy barriers for Na^+ and K^+ channels. A Na^+ channel gating barrier change of $\Delta G_h = 180$ meV was calculated from the rate constant β_h. This was possible because of the distortion due to closed-gate ion leakage across a single gating barrier. (Ch. 5, Ch. 6)

10. For sodium channels, the activation-gate time constant matches the electron-tunneling time constant at the control sites; there is negligible distortion from ion transport. For potassium channels, the electron tunneling time constant is substantially increased by the voltage dependency of ion transport across the gating-cavity energy barriers. This is represented by γ-distortion factors. (Ch. 5)

11. The random characteristic of single-channel current pulses results from random tunneling and thermal activation. The energy from the electric field needed to control individual gates is below the thermal noise energy level and the cell needs a large number of parallel ion channels to increase the signal-to-noise ratio. (Ch. 1, Ch. 5)

12. The far sites of arginine and lysine act as memory storage locations for the electron, with time constants ranging from milliseconds to minutes, hours or days, depending on tunneling distance. (Ch. 8)

13. Calcium oscillators use electron tunneling to far sites that are coupled to inactivation gates. Gating of the Ca^{2+} channel current results in regenerative feedback. With the channel open, there is an increase in the local calcium concentration and membrane potential until nearby potassium channels open and return the membrane to a negative potential. The tunneling distances and the electric field intensity across the tunneling sites control the frequency and nature of the burst mode oscillations and the resulting signaling to neurotransmitter release sites. (Ch. 8)

14. From the calcium oscillator model it is inferred that the oscillation frequencies may be increased by a hot-spot temperature rise at the NH_3 groups, resulting from microwave absorption at resonant microwave frequencies. (Ch. 8 and Ch. 10)

15. The influx gating latch-up effect can account for a number of the observations for potassium channels, including the exponent n' being greater than one, and a large attenuation for ion influx. (Ch. 7)

16. The activation gate for shaker potassium channels has a gating cavity in each of the four subunits, which is composed of five amino acid residues on the S6 α-helix. The electric field from an electron at the q_1 control site, on the S4 α-helix, closes the channel by holding a transiting ion in the gating cavity. The amino acid sequence determines the locations of activation and inactivation gating control sites on S4. These control site locations establish constraints, which define the locations of activation and inactivation gating cavities on the S6. (Ch. 9)

17. The central part of the channel in the region of the selectivity filter is closed to the passage of water and ions by the pore loop. This is necessary to maintain the large potential drop across the selectivity filter and to have ions and water travel in the region defined by the activation-gating cavity in each subunit. (Ch. 9)

18. The gating cavities also act as the selectivity filter. Substantial discrimination against the smaller sodium ion is achieved when the sodium ion penetrates into a gating cavity a distance δ beyond the potassium ion penetration. This model for selectivity is compatible with the four incremented gating cavities needed for the electron-gating model to have n^4 open-channel probability, as in the Hodgkin-Huxley model. (Ch. 9)

19. The strong repulsive force between the tunnel-track electrons can cause electrons at far sites (controlling inactivation) to couple back and influence the kinetics of activation and conversely, electron probability at activation control sites can influence the kinetics of inactivation. The tunnel-track electrons make the ion channel a sensitive interacting electrostatic system. (Ch 9)

20. Action potentials and other depolarizations of ion channels increase the probability for gating electrons to become immobilized at far sites with long time constants. Recovery requires a rest period at a hyperpolarizing potential. Both domain I and II of L-type calcium channels have far sites (q_{F17}) with a tunneling distance of ~26 Å. This distance was estimated to have a peak time constant of roughly 40 days at 6°C (Table 8-1). At 37°C and with an electric field crossing the sites, the electron-tunneling time to this site could be a day or more. All of the intervening sites are conserved and it seems likely that there would be an associated inactivation gate. When the electron tunnels to this site, the channel would then be inactivated. Recovery from this condition would require a long rest period at a hyperpolarized membrane potential. It is thought that these long time-constant inactivation gates in L-type calcium channels could be associated with the sleep cycle. (Ch. 8)

Most published research has been at the molecular level, but future research at a lower level, the level of the electron, may provide solutions to many current enigmas. Introducing the electron into gating brought constraints, which helped define the electron-gating model and indicated that nature is using the electron not only for gating, but also as an agent for signaling, timing and memory. Further understanding of the quantum-mechanical mechanisms and the role of the electron will likely bring many benefits to human health and well-being.

REFERENCES

Adrian, R. H., W. K. Chandler and A. L. Hodgkin. 1970. Slow changes in potassium permeability in skeletal muscle. *J. Physiol. (Lond.)* 208: 645-668.

Aidley, D. J. and P. R. Stanfield. 1996. *Ion Channels*, Cambridge University Press.

Alberts, B., D. Bray, J. Lewis, M. Raff, K. Roberts and J. D. Watson (eds.). 1994. *Molecular Biology of the Cell*, 3rd Ed. Garland Publishing.

Aldrich, R. W., D. P. Corey and C. F. Stevens. 1983. A reinterpretation of mammalian sodium channel gating based on single channel recording. *Nature* 306: 436-441.

Andalib, P., J. F. Consiglio, J. G. Trapani and S. J. Korn. 2004. The external TEA binding site and C-type inactivation in voltage-gated potassium channels. *Biophys. J.* 87: 3148-3161.

Armstrong, C. M. and F. Bezanilla. 1974. Charge movement associated with the opening and closing of the activation gates of the Na channels. *J. Gen. Physiol.* 63: 533-552.

Armstrong, C. M. and F. Bezanilla. 1977. Inactivation of the sodium channel, *II*. Gating current experiments. *J. Gen. Physiol.* 70: 567-590.

Ashcroft, F. M. 2000. *Ion Channels and Disease: Channelopathies.* Academic Press.

Barlow, D. J. and J. M. Thornton. 1988. Helix geometry in proteins. *J. Mol. Biol.* 201: 601-619.

Barth, A. 2000. The infrared absorption of amino acid side chains. *Prog. Biophys. Mol. Biol.* 74: 141-173.

Bean, B. P. and I. M. Mintz. 1994. Pharmacology of different types of calcium channels in rat neurons. *Handbook of Membrane Channels: Molecular and cellular Physiology*. C. Peracchia (ed.). 199-210. Academic Press.

Begenisich, T., and P. De Weer. 1980. Potassium flux ratio in voltage-clamped squid giant axons. *J. Gen. Physiol.* 76: 83-98.

Begenisich, T. 1994. Permeation properties of cloned K^+ channels. *Handbook of Membrane Channels: Molecular and cellular Physiology*. C. Peracchia (ed.). 17-28. Academic Press.

Beratan, D. N., J. N. Onuchic, J. R. Winkler and H. B. Gray. 1992. Electron-tunneling pathways in proteins. *Science* 258: 1740-1741.

Bezanilla, F. and C. M. Armstrong. 1977. Inactivation of the sodium channel. I. Sodium current experiments. *J. Gen. Physiol.* 70: 549-566.

Binstock, L. and L. Goldman. 1971. Rectification in instantaneous potassium current-voltage relations in Myxicola giant axons. *J. Physiol. (Lond.)* 217: 517-531.

Butler, A., A. G. Wei, K. Baker, and L. Salkoff. 1989. A family of putative potassium channel genes in *Drosophila*. *Science* 243: 943-947.

Catterall, W. A. 2000. Structure and regulation of voltage-gated Ca^{2+} channels. *Annu. Rev. Cell Dev. Biol.* 16: 521-555.

Chalfie, M., Y. Tu, G. Euskirchen, W. W. Ward and D. C. Prasher. 1994. Green fluorescent protein as a marker for gene expression. *Science* 263: 802-805.

Choi, K. L., C. Mossman, J. Aube and G. Yellen. 1993. The internal quaternary ammonium receptor site of *Shaker* potassium channels. *Neuron* 10: 533-541.

DeFelice, L. J. 1993. Gating currents. Machinery behind the molecule. *Biophys J.* 64: 5-6.

DeVault, D. 1984. *Quantum-mechanical tunnelling in biological systems*, 2nd Ed. Cambridge University Press.

Ehrenstein, G. and D. L. Gilbert. 1966. Slow changes of potassium permeability in the squid giant axon. *Biophys. J.* 6: 553-566.

Esaki, L. 1974. Long journey into tunneling. *Science* 183: 1149-1155.

Feynman, R. P. 1965. *The Feynman Lectures on Physics* Vol. III. *Quantum Mechanics*. Addison-Wesley Publishing Company 9: 1-15

Frankenhaeuser, B. and L. E. Moore. 1963. The effect of temperature on the sodium and potassium permeability changes in myelinated nerve fibres of *Xenopus laevis*. *J. Physiol. (Lond.)* 169: 431-437.

Galvanovskis, J. and J. Sandblom. 1997. Amplification of electromagnetic signals by ion channels. *Biophys. J.* 73: 3056-3065.

Gilly, W. F. and C. M. Armstrong. 1980. Gating current and potassium channels in the giant axon of the squid. *Biophys. J.* 29: 485-492.

Goldin, A. L. 1994. Molecular analysis of sodium channel inactivation. *Handbook of Membrane Channels: Molecular and cellular Physiology.* C. Peracchia (ed.). 121-135. Academic Press.

Goldman, D. E. 1943. Potential, impedance, and rectification in membranes. *J. Gen. Physiol.* 27: 37-60.

Gray, H. B., J. R. Winkler. 2003. Electron tunneling through proteins. *Q. Rev. of Biophys.* 36: 341-372.

Harks, E. G., J. J. Torres, L. N. Cornelisse, D. L. Ypey, A. P. Theuvenet. 2003. Ionic basis for excitability of normal rat kidney (NRK) fibroblasts. *J. Cell Physiol.* 196: 493-503.

Haupts, U., S. Maiti, P. Schwille and W. W. Webb. 1998. Dynamics of fluorescence fluctuations in green fluorescent protein observed by fluorescence correlation spectroscopy. *Proc. Natl. Acad. Sci. USA* 95: 13573-13578.

Heginbotham, L and R. MacKinnon. 1992. The aromatic binding site for tetraethylammonium ion on potassium channels. Neuron. 8: 483-491.

Heim, R., D. C. Prasher and R. Y. Tsien. 1994. Wavelength mutations and posttranslational autoxidation of green fluorescent protein. *Proc. Natl. Acad. Sci. USA* 91: 12501-12504.

Hille, B. 1975. Ionic selectivity, saturation and block in sodium channels. A four-barrier model. *J. Gen. Physiol.* 66: 535-560.

Hille, B. 2001. *Ionic Channels of Excitable Membranes*, 3rd Ed. Sinauer Associates, Inc.

Hirschberg, B., A. Rovner, M. Lieberman and J. Patlak. 1995. Transfer of twelve charges is needed to open skeletal muscle Na^+ channels. *J. Gen. Physiol.* 106: 1053-1068.

Hodgkin, A. L. and A. F. Huxley. 1952. A quantitative description of membrane current and its application to conduction and excitation in nerve. *J. Physiol.* 117: 500-544.

Hodgkin, A. L. and R. D. Keynes. 1955. The potassium permeability of a giant nerve fiber. *J. Physiol.* 128: 61-88.

Hogan, K., P. A. Powers and R. G. Gregg. 1994. Cloning of the human skeletal muscle α_1 subunit of the dihydropyridine-sensitive L-type calcium channel (CACNL1A3). *Genomics* 24: 608-609.

Hopfield, J. J. 1974. Electron transfer between biological molecules by thermally activated tunneling. *Proc. Natl. Acad. Sci. USA* 71:3640-3644.

Hoshi, T., W. N. Zagotta and R. W. Aldrich. 1990. Biophysical and molecular mechanisms of *Shaker* potassium channel inactivation. *Science* 250: 533-538.

Hoshi, T., W. N. Zagotta and R. W. Aldrich. 1991. Two types of inactivation in *Shaker* K$^+$ channels: Effects of alterations in the carboxy-terminal region. *Neuron* 7: 547-556.

Hutson, M. S., U. Alexiev, S. V. Shilov, K. J. Wise and M. S. Braiman. 2000. Evidence for a perturbation of arginine-82 in the bacteriorhodopsin photocycle from time-resolved infrared spectra. *Biochemistry* 39 :13189-13200.

Johnston, D. and S. M. Wu. 1995. *Foundations of Cellular Neurophysiology*, MIT Press, Cambridge, MA.

Kandel, E. R., J. H. Schwartz and T. M. Jessell (eds.). 2000. *Principles of Neural Science*, 4th Ed. McGraw-Hill.

Keynes, R. D. and E. Rojas. 1974. Kinetics and steady-state properties of the charged system controlling sodium conductance in the squid giant axon. *J. Physiol. (Lond.)* 239: 393-434.

Koch, W. J., P. T. Ellinor and A. Schwartz. 1990. cDNA cloning of a dihydropyridine-sensitive calcium channel from rat aorta. Evidence for the existence of alternatively spliced forms. *J. Biol. Chem.* 265: 17786-17791.

Langen, R., I. J. Chang, J. P. Germanas, J. H. Richards, J. R. Winkler, H. B. Gray. 1995. Electron tunneling in proteins: coupling through a β strand. *Science* 268: 1733-1735.

Lee, A. L., K. A. Sharp, J. K. Kranz, X. J. Song and A. J. Wand. 2002. Temperature dependence of the internal dynamics of a calmodulin-peptide complex. *Biochemistry.* 41: 13814-13825.

Levitan, I. B., L. K. Kaczmarek. 2002. *The Neuron*, 3rd Ed. Oxford University Press

Liu, Y., M. Holmgren, M. E. Jurman and G.Yellen. 1997. Gated access to the pore of a voltage-dependent K$^+$ channel. *Neuron* 19: 175-184.

Lodish, H., A. Berk, A. L. Zipursky, P. Matsudaira, D. Baltimore and J. Darnell. 2000. *Molecular Cell Biology*, 4th Ed. W. H. Freeman, New York.

Loeser, J. G., C. A. Schmuttenmaer, R. C. Cohen, M. J. Elrod, D. W. Steyert, R. J. Saykally, R. E. Bumgarner and G. A. Blake. 1992. Multidimensional hydrogen tunneling dynamics in the ground vibrational state of the ammonia dimer. *J. Chem. Phys.* 97(7): 4727-4749.

Ma, W. J., R. W. Holz and M. D. Uhler. 1992. Expression of a cDNA for a neuronal calcium channel alpha 1 subunit enhances secretion from adrenal chromaffin cells. *J. Biol. Chem.* 267: 22728-22732.

MacKinnon, R. and G. Yellen. 1990. Mutations affecting TEA blockade and ion permeation in voltage-activated K^+ channels. *Science* 250: 276-279.

Manning, M. F. 1935. Energy levels of a symmetrical double minima problem with applications to the NH_3 and ND_3 molecules. *J. Chem. Phys.* 3:136-138.

Mannuzzu, L., M. M. Moronne, and E. Y. Isacoff. 1996. Direct physical measure of conformational rearrangement underlying potassium channel gating. *Science* 271: 213-216.

Marcus, R. A. and N. Sutin. 1985. Electron transfers in chemistry and biology. *Biochim. Biophys. Acta* 811: 265-322.

Melishchuk, A. and C. M. Armstrong. 2001. Mechanism underlying slow kinetics of the OFF gating current in *Shaker* potassium channel. *Biophys. J.* 80: 2167-2175.

Mikami, A., K. Imoto, T. Tanabe, T. Niidome, Y. Mori, H. Takeshima, S. Narumiya and S. Numa. 1989. Primary structure and functional expression of the cardiac dihydropyridine-sensitive calcium channel. *Nature* 340:230-233.

Moser, C. C., J. M. Keske, K. Warncke, R. S. Farid and P. L. Dutton. 1992. Nature of biological electron transfer. *Nature* 355: 796-802.

Neher, E. and B. Sakmann. 1976. Single-channel currents recorded from membrane of denervated frog muscle fibres. *Nature* 260: 799-802.

Nicholls, J. G., R. A. Martin, B. G. Wallace and P. A. Fuchs. 2001. *From Neuron to Brain,* 4th Ed. Sinauer Associates, Inc., Sunderland, MA

Noda, M., S. Shimizu, T. Tanabe, T. Takai, T. Kayano, T. Ikeda, H. Takahashi, H. Nakayama, Y. Kanaoka, N. Minamino, K. Kangawa, H. Matsuo, M. A. Rafterty, T. Hirose, S. Inayama, H. Hayashida, T. Miyata and S. Numa. 1984. Primary structure of *Electrophorus electricus* sodium channel deduced from cDNA sequence. *Nature* 312: 121-127.

Nonner, W. 1980. Relations between the inactivation of sodium channels and the immobilization of gating charge in frog myelinated nerve. *J. Physiol. (Lond.)* 299: 573-603.

Ormö, M., A. B. Cubitt, K. Kallio, L. A. Gross, R. Y. Tsien and S. J. Remington. 1996. Crystal structure of the *Aequorea victoria* green fluorescent protein. *Science* 273: 1392-1395.

Page, C. C., C. C. Moser, X. Chen and P. L. Dutton. 1999. Natural engineering principles of electron tunneling in biological oxidation–reduction. *Nature* 402: 47-52.

Patlak, J. 1991. Molecular kinetics of voltage-dependent Na^+ channels. *Physiol. Rev.* 71: 1047-1079.

Pauling, L., R. B. Corey and H. R. Branson. 1951. The structure of proteins: two hydrogen-bonded helical configurations of the polypeptide chain. *Proc. Natl. Acad. Sci. USA* 37: 205-211.

Pongs, O., N. Kecskemethy, R. Müller, I. Krah-Jentgens, A. Baumann, H. H. Kiltz, I. Canal, S. Llamazares and A. Ferrus. 1988. *Shaker* encodes a family of putative potassium channel proteins in the nervous system of *Drosophila*. *EMBO J.* 7: 1087-1096.

Rettig, J., S. H. Heinemann, F. Wunder, C. Lorra, D. N. Parcej, O. J. Dolly and O. Pongs. 1994. Inactivation properties of voltage-gated K^+ channels altered by presence of β-subunit. *Nature* 369: 289-294.

Rosenthal, J. J. C. and W. F. Gilly. 1993. Amino acid sequence of a putative sodium channel expressed in the giant axon of the squid *Loligo opalescens*. *Proc. Natl. Acad. Sci. USA*. 90: 10026-10030. NCBI Accession L19979.

Rosenthal, J. J. C., R. Vickery and W. F. Gilly. 1996. Molecular identification of SqKv1A: A candidate for the delayed rectifier K channel in squid giant axon. *J. Gen.Physiol.* 108: 207-219. NCBI Accession U50543.

Ruppersberg, J. P., R. Frank, O. Pongs and M. Stocker. 1991. Cloned neuronal $I_K(A)$ channels reopen during recovery from inactivation. *Nature* 353: 657-660.

Salkoff, L., A. Butler, A. Wei, N. Scavarda, K. Griffen, C. Ifune, R. Goodman and G. Mandel. 1987. Genomic organization and deduced amino acid sequence of a putative sodium channel gene in *Drosophila*. *Science* 237: 744-749.

Saykally, R. J. and G. A. Blake. 1993. Molecular interactions and hydrogen bond tunneling dynamics: some new perspectives. *Science* 259: 1570-1575.

Sigworth, F. J. 1993. Voltage gating of ion channels. *Q. Rev. of Biophys.* 27: 1-40.

Sisler, H. H., 2002. Liquid ammonia as a solvent. *Encyclopedia of Science and Technology* 9th Ed. Mc Graw-Hill. Vol. 1: 565-566.

Soldatov, N. M. 1992. Molecular diversity of L-type Ca^{2+} channel transcripts in human fibroblasts. *Proc. Natl. Acad. Sci. U.S.A.* 89: 4628-4632.

Tanabe, T., H. Takeshima, A. Mikami, V. Flockerzi, H. Takahashi, K. Kangawa, M. Kojima., H. Matsuo, T. Hirose and S. Numa. 1987. Primary structure of the receptor for calcium channel blockers from skeletal muscle. *Nature* 328: 313-318.

Tempel, B. L., D. M. Papazian, T. L. Schwarz, Y. N. Jan and L. Y. Jan. 1987. Sequence of a probable potassium channel component encoded at *Shaker* locus of *Drosophila*. *Science* 237: 770-775.

Townes, C. H. and A. L. Schawlow. 1975. *Microwave Spectroscopy.* Published 1955. McGraw-Hill Book Company. Republished with corrections 1975. Dover Publications, Inc.

Unwin, N. 1989. The structure of ion channels in membranes of excitable cells. *Neuron* 3: 665-676.

Ussing, H. H. 1949. The distinction by means of tracers between active transport and diffusion. *Acta Physiol. Scand.* 19: 43-56.

Vandenberg, C. A. and F. Bezanilla. 1991. Single channel, macroscopic, and gating currents from sodium channels in the squid giant axon. *Biophys. J.* 60: 1499-1510.

Wei, A., M. Covarrubias, A. Butler, K. Baker, M. Pak and L Salkoff. 1990. K^+ Current diversity is produced by an extended gene family conserved in *Drosophila* and mouse. *Science* 248: 599-603.

Winkler, J. R., A. J. Di Bilio, N. A. Farrow, J. H. Richards and H. B. Gray. 1999. Electron tunneling in biological molecules. *Pure Appl. Chem.* 71: 1753-1764.

Woodhull, A. M. 1973. Ionic blockage of sodium channels in nerve. *J. Gen. Physiol.* 61: 687-708.

Yang, F., L. G. Moss and G. N. Phillips, Jr. 1996. The molecular structure of green fluorescent protein. *Nature Biotechnol.* 14: 1246-1251.

Yellen, G., M. E. Jurman, T. Abramson and R. MacKinnon. 1991. Mutations affecting internal TEA blockade identify the probable pore-forming region of a K^+ channel. *Science* 251: 939-942.

Yool, A. J. and T. L. Schwarz. 1991. Alteration of ionic selectivity of a K^+ channel by mutation of the H5 region. *Nature* 349: 700-704.

Zagotta, W. N., T. Hoshi and R. Aldrich. 1990. Restoration of inactivation in mutants of *Shaker* potassium channels by a peptide derived from ShB. *Science* 250: 568-571.

Zhou, J., L. Cribbs, J. Yi, R. Shirokov, E. Perez-Reyes and E. Rios. 1998. Molecular cloning and functional expression of a skeletal muscle dihydropyridine receptor from *Rana catesbeiana. J. Biol. Chem.* 273: 25503-25509.

Zwart, E., H. Linnartz, W. L. Meerts, G. T. Fraser, D. D. Nelson and W. Klemperer. 1991. Microwave and submillimeter spectroscopy of $Ar-NH_3$ states correlating with $Ar + NH_3$ ($j=1, |k|=1$). *J. Chem. Phys.* 95: 793-803.

INDEX

Activation gate. *See gating and distortion factors*
Amino acid sequences
  K_v channels, 131-132, 145
  L-type calcium channel, 124
Ammonia. *See NH₃ groups*
Amplification
  and negative conductance, 8-10
  and the inverted region, 51-55
  energy window, 30
  equations, 17, 21-22
  molecular stretching model, 15-21
  selectivity curve, 28
  sensitivity, 8, 17-18
  theory, 21-26
  voltage dependent, 26-29
Arginine thermal model, 159
Atomic mass unit, 11
Avogadro constant, 11

Back sites, 117
  and hyperpolarization, 117-118
Bacteriorhodopsin, 167-170
  photocycle perturbation
    at R82, 167
  WT-R82A spectra, 167-169
Basic amplification defined, 27
Blue fluorescent protein, 152-165
  fluorophore, 152
  microwave spectra, 153-156
  sampling method, 152-153

Boltzmann constant, 11

Calcium oscillator, 123-128
  bursting pacemaker, 127
  burst mode, 125-127
  ELF magnetic fields, 127
  far-site model, 125
  frequency adaptation, 127
  hypoPP mutations, 128
  L-type S4 sequence, 124
  microwave sensitivity, 166
  regenerative feedback, 127
  sleep mode, 128
Capacitance factor, 11, 40
Charge immobilization, 122-123
Coupling between
  tunnel track electrons, 142-144
Current-voltage relationships, 9

Delayed rectifier
  See potassium channels
Differential mode-probability, 27-28
Dipole moment of NH₃, 24
Displacement energy ratio, 11
Drosophila
  See Shaker potassium channels
Dwell time, 62

Edge distortion of gating, 81-82
Electric field distance for Na⁺
  channel, 11

187

Printed in the USA
CPSIA information can be obtained
at www.ICGtesting.com
JSHW011506221024
72173JS00005B/1219